Computer Simulated Experiments for Digital Electronics Using Electronics Workbench Multisim®

Second Edition

Richard H. Berube
Community College of Rhode Island

PEARSON
Prentice
Hall

Prentice Hall

Upper Saddle River, New Jersey
Columbus, Ohio

Acquisitions Editor: Dennis Williams
Development Editor: Kate Linsner
Production Editor: Rex Davidson
Design Coordinator: Diane Ernsberger
Cover Designer: Jeff Vanik
Production Manager: Matt Ottenweller
Marketing Manager: Ben Leonard

This book was printed and bound by Courier Kendallville, Inc. The cover was printed by Coral Graphic Services, Inc.

Pearson Education Ltd.
Pearson Education Singapore Pte. Ltd.
Pearson Education Canada, Ltd.
Pearson Education—Japan

Pearson Education Australia Pty. Limited
Pearson Education North Asia Ltd.
Pearson Educación de Mexico, S.A. de C.V.
Pearson Education Malaysia Pte. Ltd.

10 9 8 7 6 5 4 3 2 1

ISBN: 0-13-048785-6

Preface

Computer Simulated Experiments for Digital Electronics Using Electronics Workbench Multisim®, Second Edition, is a unique and innovative book that uses Multisim to simulate digital laboratory experiments on a computer. Computer simulated experiments do not require extensive laboratory facilities, and a computer provides a safe and cost-effective laboratory environment. The digital circuits can be modified easily with on-screen editing, and analysis results provide faster and better feedback than a series of experiments using hardwired digital circuits. The materials list and circuit diagrams included with each experiment make this book also usable in a hardwired laboratory environment, if hardwired experience is desired.

The experiments in this book are designed to help reinforce the theory learned in a digital electronics course. By answering questions about the results of each experiment, students will develop a clearer understanding of the theory. Also, the interactive nature of these experiments encourages student participation, which leads to more effective learning and a longer retention of the digital concepts.

The experiments in Part I of this book involve the study of logic gates and combinational logic circuits. The experiments in Part II involve the study of arithmetic logic circuits such as binary adders, BCD adders, parity generators and checkers, and magnitude comparators. The experiments in Part III involve medium scale integrated (MSI) circuits such as decoders, encoders, multiplexers, and demultiplexers. The experiments in Part IV involve the study of sequential logic circuits such as latches, flip-flops, monostable and astable multivibrators, registers, and counters. The experiments in Part V involve the study of circuits that interface the digital world with the analog world for the acquisition of data. A series of troubleshooting problems is included at the end of each section to help students develop troubleshooting skills. In each troubleshooting problem, the parts bin has been removed to force the student to find the fault by making a series of circuit measurements rather than by replacing components. A solutions manual showing measured data, answers to the questions, and answers to the troubleshooting problems is available to instructors.

In the second edition, the circuits have been modified and some of the experiments have been changed so that they will work better with the new Electronics Workbench Multisim circuit simulator. Experiments 6, 7, and 8 were modified so that the computer circuits are the same as they would be in a hardwired laboratory. The name of the Preparation section of each experiment has been changed to a Theory section. This Theory section has all of the technical information needed to do the calculations and answer the questions in the Procedure section without referring to another textbook, and makes it possible to use this book as a combination text and lab manual, if desired. These modifications should make these experiments work better with the new Multisim circuit simulator.

Richard H. Berube
Email: rberube@ccri.edu

Acknowledgments

The author appreciates the dedication and talent of the editorial staff of Prentice Hall, particularly Dennis Williams and Rex Davidson for their encouragement to complete this second edition. I am also grateful to my copy editor, Ben Shriver, whose attention to detail on a short time schedule was extremely valuable.

R. H. Berube
email rberube@ccri.edu

Contents

Introduction

Electronics Workbench Multisim® is similar to a real electronics laboratory environment, except that circuits are simulated on a computer and results are obtained more quickly. All of the components and instruments necessary to create and simulate mixed mode analog and digital circuits on a computer are provided. Using a mouse, you can build a circuit in the central workspace on the computer screen, attach simulated test instruments, simulate actual circuit performance, and display the results on the test instruments. Because **circuit faults can be introduced without destroying or damaging actual components,** more extensive troubleshooting experiments can be performed using Electronics Workbench Multisim. Also, faulty components that are deliberately introduced in a circuit simulated on a computer can help make it easier to find the faulty component in an actual circuit. The **Multisim Help menu** has all of the information needed to get started using Multisim. **Additional notes on using Multisim** have been included in **Appendix B** at the end of this book.

Each experiment includes a list of **Objectives, Materials list, Theory section, circuit diagrams,** and a **Procedure section.** The Procedure section requires you to record measured data, draw logic circuits, write logic equations, calculate expected values, and answer a series of questions designed to reinforce the theory. The Theory section provides all of the theory needed to complete the procedure and answer the questions. **The Materials list and circuit diagrams make it possible to use this manual in a hardwired laboratory environment.** A series of **troubleshooting problems** is included at the end of each section.

If this book is used in a hardwired laboratory, wire the circuit from the circuit diagram and connect the instruments specified. The **pin diagrams for the IC chips** required for each experiment are shown in **Appendix A.** Don't forget to connect each IC chip V_{CC} **terminal to +5 V** and each **GND terminal** to the power supply ground. After the digital circuit is wired and checked, turn on the power, record the data in the space provided, and answer the questions. You also can **wire the circuits on the computer screen yourself** by selecting the IC chips from the Multisim parts bin. By wiring the IC chips in the Multisim workspace, you can **create a hardwired laboratory environment.**

The **CD-ROM** provided with this text contains the digital circuits needed to perform the experiments, and the troubleshooting circuits. These circuits require that you have access to Multisim 7 in your school lab (the Lab Edition) or on your computer (Electronics Workbench Student Suite). If you do not currently have access to this software and wish to purchase it, please call Prentice Hall Customer Service at 1-800-282-0693 or send a fax request to 1-800-835-5327.

If you need technical assistance or have questions concerning the Multisim 7 software, contact Electronics Workbench directly for support at (416) 977-5550 or visit the EWB website at www.electronicsworkbench.com.

PART

I Combinational Logic Circuits

The experiments in Part I involve the study of **logic gates** and **combinational logic circuits.** A combinational logic circuit consists of a combination of logic gates that produces a logic circuit. In combinational logic circuits, the output response follows changes in the input with **minimum delay**. Prior input conditions have no effect on the present output levels because combinational logic circuits do not have memory.

The logic gates you will study are INVERTERS, OR gates, AND gates, NAND gates, and NOR gates. You will review the **Boolean theorems** and use them to analyze and simplify combinational logic circuits. You will learn how to use **Karnaugh maps** to help simplify and design combinational logic circuits. In the final experiment in Part I you will solve some **troubleshooting problems** in combinational logic circuits.

The logic circuits for the experiments in Part I can be found on the enclosed disk in the PART1 subdirectory.

EXPERIMENT

Preliminary Concepts

Objectives:

1. Investigate the Electronics Workbench Multisim digital word generator.
2. Investigate the Electronics Workbench Multisim logic analyzer.
3. Lean how to use a logic probe to determine logic levels.
4. Determine the voltage levels that represent a binary "one" and a binary "zero."

Materials:

One digital word generator (Multisim only)
One logic analyzer
One 5 V dc voltage supply
One dc voltmeter
One logic probe
One logic switch
Resistors—500 Ω, 1 kΩ, 2 kΩ, 10 kΩ

Theory:

Word Generator (Multisim V6)

Open circuit file FIG1-1 and double click the word generator to bring down the enlargement. The Electronics Workbench Multisim **word generator** stores **32-bit binary words.** The 32-bit words stored are listed as 8-character hexadecimal numbers in the left column scroll box. If you want to change a 32-bit word stored in the word generator, you must first select the word in the left column scroll box. Then type an 8-character hexadecimal number in the box labeled HEX, type a 32-bit number in the box labeled BINARY, or type the equivalent ASCII character in the box labeled ASCII. (This will change the selected hexadecimal number in the left column scroll box). You can select the HEX box, the BINARY box, or the ASCII box using the mouse arrow and clicking the left mouse button. When a word in the left scroll box is selected, its hexadecimal word address appears in the EDIT box.

The 32-bit binary words are **transmitted** at the 32 output terminals by clicking the STEP, BURST, or CYCLE buttons on the word generator. To transmit one word at a time, click the STEP button. To transmit a group of words in sequence, click the BURST button. To transmit a continuous stream of words, click the CYCLE button. To stop the transmission of the continuous stream of words, click the On-Off switch. The word being transmitted is highlighted in the left column scroll box. You can select the **first word** to be transmitted by typing the hexadecimal word address in the box labeled INITIAL. You can select the **last word** to be transmitted by typing the hexadecimal word address in the box labeled FINAL. As the word generator outputs words, each hexadecimal **word address** appears in the box labeled CURRENT.

Each word is transmitted for the duration of **one clock period (T)** of the internal clock when INTERNAL trigger is selected. The **clock frequency (f)** can be selected by changing the frequency selection on the front of the word generator. When EXTERNAL trigger is selected, the clock pulse input at the external trigger terminal is used as the clock. The word generator can be **triggered** on the **ascending** or **descending edge** of a clock pulse by selecting the appropriate trigger selection.

The Multisim word generator has a **breakpoint** feature that allows you to set breakpoints. Use a breakpoint when you want to "pause" a stream of words at a specific word. To insert a breakpoint, select a word in the left scroll window, and then click the BREAKPOINT box on the word generator. (An asterisk will appear next to the word). To remove the breakpoint, click on an existing breakpoint in the scroll window, and then click the BREAKPOINT box. More than one breakpoint can be set. Breakpoints can be used in both the **burst** and **cycle modes.**

For more information about the Multisim word generator, select HELP to bring down the help screen, select INSTRUMENTS, and then select WORD GENERATOR.

Word Generator (Multisim V7)

Open circuit file FIG1-1 and double click the word generator to bring down the enlargement. The Electronics Workbench Multisim **word generator** stores **32-bit binary words.** The 32-bit binary words stored are listed in the right column scroll box. You can display the 32-bit words in hexadecimal, decimal, binary, or ASCII code by making the selection under DISPLAY. If you want to change a 32-bit word stored in the word generator, you must first select the word in the right column scroll box, and then type the new word. The number of binary words stored is determined by the **buffer size.** You can change the word generator buffer size by clicking the SET button to bring down the **settings menu.** The buffer size can be displayed in hexadecimal or decimal code. You can set the **cursor**, and the **initial** and **final positions** in the right column scroll box by clicking the right mouse button on a word in the scroll box and making a selection from the pull down menu.

The 32-bit binary words are **transmitted** at the 32 output terminals by clicking the STEP, BURST, or CYCLE buttons on the word generator. To transmit one word at a time starting at the cursor position, click the STEP button. To transmit a group of words in sequence between the initial and final positions starting at the cursor position, click the BURST button. To transmit a continuous stream of words starting at the cursor position, click the CYCLE button. To stop the transmission of the continuous stream of words, click the On-Off switch. The word being transmitted is highlighted in the right column scroll box.

Each word is transmitted for the duration of **one clock period (T)** of the internal clock when INTERNAL trigger is selected. The **clock frequency (f)** can be selected by changing the frequency selection on the front of the word generator. When EXTERNAL trigger is selected, the clock pulse input at the external trigger terminal is used as the clock. The word generator can be **triggered** on the **ascending** or **descending edge** of a clock pulse by selecting the appropriate trigger selection.

The Multisim word generator has a **breakpoint** feature that allows you to set breakpoints. Use a breakpoint when you want to "pause" a stream of words at a specific word. To insert a breakpoint, click the right mouse button on a word in the scroll box and select SET BREAKPOINT from the pull down menu. (A stop sign will appear next to the word.) To remove the breakpoint, click the right mouse button on the existing breakpoint in the scroll window and select DELETE BREAKPOINT from the pull down menu. (The stop sign next to the word will disappear.) More than one breakpoint can be set. Breakpoints can be used in both the **burst** and **cycle modes.**

For more information about the Multisim word generator, select HELP to bring down the help screen, select INSTRUMENTS, and then select WORD GENERATOR.

Logic Analyzer Theory

A **logic analyzer** is used for **acquisition** and **display** of **logic states** for **timing analysis** of digital systems. Although a logic analyzer is **similar to a multichannel oscilloscope**, it displays only **multichannel digital data**. It samples the state of several channels of rapidly changing digital logic levels on the selected transition of each clock pulse. These **logic levels** are **stored** in a memory section. The clock pulses can be generated internally or externally. The clock frequency should be higher than the frequency of the digital data so that data will not be missed.

A **trigger event** determines the input data displayed. A logic analyzer can be triggered on the selected transition of an input pulse, the selected transition of the signal applied to the external trigger input, or a particular input binary word. A logic analyzer **stores data samples** until it reaches the number of samples determined by the **pre-trigger setting**. It then begins discarding old samples as new samples appear, until it sees the **trigger event**. After the trigger event, **samples are stored** until they reach the number of samples determined by the **post-trigger setting**.

After the **trigger event**, the stored logic levels are **continuously displayed** on a CRT screen as **multitrace square waves** with the **trigger point marked**. Because of their storage capability, logic analyzers display digital data before and after the trigger event. The pre-trigger multitrace data has already occurred and is **not in real-time** like the waveshapes displayed on an oscilloscope screen.

Logic Analyzer (Multisim)

Open circuit file FIG1-1 and double click the logic analyzer to bring down the enlargement. The Multisim **logic analyzer** displays up to **sixteen digital waveshapes**. These sixteen digital waveshapes are applied at the sixteen input terminals on the left side of the instrument. Each of the input terminals corresponds to one horizontal display on the logic analyzer screen. The top row of the display is the waveshape entering the top terminal on the left, and the bottom row of the display is the waveshape entering the bottom terminal on the left. The **bottom row** displays the **most-significant bit** (bit 15) in a digital word, and the **top row** displays the **least-significant bit** (bit 0). If the digital display goes beyond the screen on the right, use the scroll bar on the bottom of the logic analyzer screen to view the remaining waveshapes. You can **clear the display** by clicking the RESET button. In Multisim 7, change the screen from **black** to **white** by clicking the REVERSE button after running a simulation.

When the circuit is activated, the logic analyzer **samples** the input values on the sixteen input terminals and **stores** the binary data samples until it reaches the **pre-trigger number** of samples. Then it begins discarding old samples as new samples appear until it sees the **trigger signal**. After the trigger signal, samples are stored up to the **post-trigger number** of samples. When the trigger signal is seen, the logic analyzer **displays the pre- and post-trigger data** on the screen. To dump the stored data to the screen when the logic analyzer is not triggered, click the STOP button. If the logic analyzer is already triggered and displaying data, STOP has no effect.

To specify the **number of samples** stored **before and after triggering**, click the SET button in the **Clock box** on the logic analyzer. The **Clock Setup dialog box** will appear and display the pre-trigger and post-trigger samples. These values can be changed in the dialog box by **selecting** them with the **mouse arrow** and **clicking** the **left mouse button**. Click ACCEPT to accept the changes.

The logic analyzer can be made to **trigger** upon reading a **specified word** or **combination of words**. To specify the **trigger word** or **word combination**, click the SET button in the **Trigger box** on the logic analyzer. The **Trigger Settings dialog box** will appear. You can specify up to three trigger words or word combinations. An x specifies a don't care value. A pattern of all x's will cause the logic analyzer to trigger on the first binary word input. The **trigger qualifier** is an input signal that filters the triggering signal. If x is selected, then the qualifier is disabled and the trigger signal determines when the logic analyzer is triggered. If it is set to 1 or 0, the logic analyzer is triggered only when the trigger signal matches the qualifier level.

The clock informs the logic analyzer when to read input samples. The samples can be read on the **positive edge** or the **negative edge** of the clock pulses. The **clock frequency** (Internal clock rate) determines the **sampling rate**. If the sampling rate is **lower** than the **input data rate**, data will be **lost**. If the sampling rate is **higher** than the **input data rate**, the pre-trigger and post-trigger samples settings must be high enough to display all of the input data because each input data bit will be sampled more than once. For example, if the clock frequency is ten times the frequency of the input data, each input data bit will be sampled ten times. When the INTERNAL clock mode is selected, the **internal clock** is in control. When the EXTERNAL clock mode is selected, an **external clock** is in control. To adjust the clock settings, click SET in the **Clock box** on the logic analyzer and the **Clock Setup dialog box** will appear. The **clock qualifier** is an input signal that filters the external clock signal. If it is set to x, then the qualifier is disabled and the clock determines when samples are read. If it is set to 1 or 0, the samples are read only when the clock level matches the selected qualifier level.

The **Threshold Voltage** setting determines the minimum voltage that will be recognized as a binary one. The threshold voltage can be changed by clicking the SET button in the **Clock box** on the logic analyzer and changing the threshold voltage value in the **Clock Setup dialog box.**

The **Clocks/Div** setting determines the **horizontal time scale**. Reducing the setting will expand the time scale.

The red and blue **cursor lines** that appear on the left and right sides of the logic analyzer screen when a simulation is started are used for measuring **time differentials** between points on the curve plots. (Make sure you click REVERSE on the logic analyzer to produce a white screen in Multisim 7 after the simulation is started.) These cursor lines can be moved by **clicking and dragging** them with the mouse arrow in the same way that they are moved on the oscilloscope. The cursor time points can be read in the box at the bottom of the logic analyzer.

For more detailed information about the Multisim logic analyzer, select HELP to bring down the HELP screen, select INSTRUMENTS, and select LOGIC ANALYZER.

Experiment 1

Logic Voltage Levels

TTL logic gates recognize only **high** and **low input voltages** and produce only **high** and **low output voltages.** A voltage between 2 V and 5 V is recognized as a **logical high (1)** and a voltage between 0 V and 0.8 V is recognized as a **logical low (0)** by TTL logic device inputs. Voltages between 0.8 V and 2 V cannot be interpreted by TTL logic devices.

A **logic probe** is used to detect a binary high (1) or a binary low (0) at various terminals in a digital logic circuit. In the Multisim logic probe, a light is displayed when a binary high (1) is present (voltage above 2 V) and the light will go out when a binary low (0) is present (voltage below 0.8 V). A **hardwired logic probe** will normally have **three lights**, one for detecting a **binary high**, one for detecting a **binary low**, and one for detecting an **open circuit** or a voltage that is **out-of-range.**

A **logic switch** is used to input a binary high (2–5 V) or a binary low (0–0.8 V) to a logic circuit, as shown in Figure 1-2. The circuit in Figure 1-2 will demonstrate how logic voltage levels are measured using a logic probe and how a logic voltage level can be changed using a logic switch.

Figure 1-1 Word Generator and Logic Analyzer (Multisim)

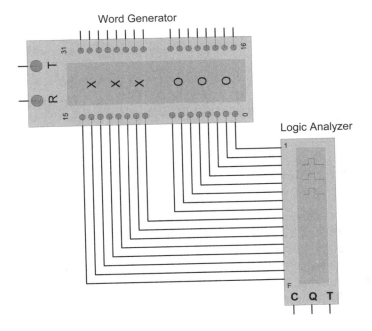

Figure 1-2   Logic Voltage Levels

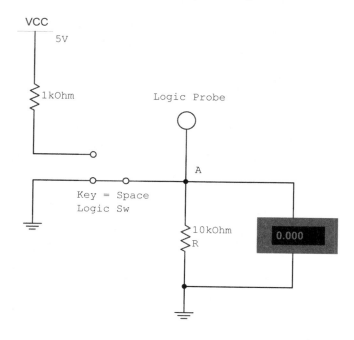

NOTE: If you are using this manual in a hardwired laboratory, see your lab instructor for information on the instruments and equipment you will be using. Because a word generator may not be available, you may need to start this experiment at the Logic Voltage Levels section of the procedure.

Word Generator and Logic Analyzer (Multisim)

Step 1.   Open circuit file FIG1-1. Bring down the word generator enlargement and make sure that the following settings are selected: Frequency = 1 kHz, Trigger = Internal, Hex codes (00000000, 00001111, 00002222, 00003333 through 0000FFFF, repeated twice). Move the word generator to the left by clicking and dragging. Bring down the logic analyzer enlargement and make sure that the following settings are selected: Clocks/Div = 32, Clock Setup (Clock Source = Internal, Clock Rate = 10 kHz, Pre-trigger Samples = 160, Post-trigger Samples = 160, Threshold Voltage = 2.5), Trigger Settings (Trigger Clock Edge = Positive, Trigger Qualifier = x, Pattern A= xxxxxxxxxxxxxxxx, Trigger Combinations = A). Click STEP on the word generator to start the simulation. Notice that the first 16-bit binary word has been transmitted and displayed on the logic analyzer screen from bottom-to-top.

Questions: Is the 16-bit binary word transmitted by the word generator the same 16-bit word displayed on the logic analyzer screen from bottom-to-top? What is the binary word?

FFFF

Step 2.   Click STEP on the word generator again and notice that the second 16-bit binary word has been transmitted and displayed on the logic analyzer screen from bottom-to-top.

Questions: Is the second 16-bit binary word transmitted by the word generator the same 16-bit word displayed on the logic analyzer screen from bottom-to-top? What is the binary word?

 EEEE

Step 3.   Continue to click STEP on the word generator until sixteen of the 16-bit binary words have been transmitted and displayed on the logic analyzer screen.

Questions: Are all sixteen of the transmitted binary words the same words displayed on the logic screen from bottom-to-top?

 YES

Why is each binary word displayed from bottom-to-top on the logic analyzer screen?

Step 4.   Click BURST on the word generator. Notice that the word generator transmitted 32 binary words in sequence, but the logic analyzer only displayed 16 binary words. **Note the Internal clock rate and the Post-trigger samples setting on the logic analyzer.**

Question: Why were only 16 binary words displayed on the logic analyzer screen even though 32 binary words were transmitted by the word generator?

THE PATTERN WAS SOT UP TO 16

Step 5.   Click the SET button in the Clock box on the logic analyzer. Change the Post-trigger samples to 320 and click ACCEPT. Click RESET on the logic analyzer. Click BURST on the word generator.

Questions: How many 16-bit binary words were displayed on the logic analyzer screen? Did it change from the number in Step 4? **Explain.**

YES, 32

Step 6.   Change the Clocks/Div to 64 and click RESET on the logic analyzer. Click CYCLE on the word generator. Notice that the word generator cycles through binary words continuously. Stop the simulation after the logic analyzer stops displaying data.

Questions: Did the logic analyzer continue displaying all of the pulses generated? **Explain why.**

 NO

Step 7.   Click the SET button in the Clock box on the logic analyzer. Change Pre-trigger samples to 60 and Post-trigger samples to 260 in the Clock Setup dialog box, and then click ACCEPT. Click the SET button in the Trigger box on the logic analyzer. Change the Pattern A word to 0110011001100110 (6666H), select A for Trigger combinations, and then click ACCEPT. Click CYCLE on the word generator. Notice that the logic analyzer displayed the pre-trigger data (before the green line) and post-trigger data (after the green line). Stop the simulation.

Questions: Did the logic analyzer trigger (green line) on the correct binary code (6666H)?

YES

How many post-trigger 16-bit binary words were displayed (after the green line)?

Did this match the logic analyzer settings?

Step 8.   Change the Clocks/Div setting back to 32. Click the SET button in the Clock box on the logic analyzer. Change the Post-trigger samples setting to 100, and then click ACCEPT. Click CYCLE on the word generator to start the simulation. Run the simulation until the logic analyzer stops displaying data, then stop the simulation.

Questions: Did the logic analyzer trigger (green line) on the correct binary code (6666H)?

How many post-trigger 16-bit binary words were displayed (after the green line)?

16

Explain any difference from the Step 7 results.

Step 9.   Click the SET button in the Clock box on the logic analyzer. Change the clock frequency (Internal clock rate) to 20 kHz, and then click ACCEPT. Click CYCLE on the word generator to start the simulation. Run the simulation until the logic analyzer finishes displaying data, then stop the simulation.

Questions: How many post-trigger 16-bit binary words were displayed?

Experiment 1

Explain any difference from the results in Step 8.

Logic Voltage Levels

Step 1.   Open circuit file FIG1-2. Click the On-Off switch to run the simulation. Record the voltage on terminal A (V_A).

$V_A =$ _____

Questions: Does voltage V_A represent binary "zero" or a binary "one"?

Is the logic probe light on or off? Does this indicate a binary "zero" or a binary "one"?

Step 2.   Press the space bar on the computer keyboard to change the logic switch to the up position. After the voltmeter changes to the new reading, stop the simulation. Record the voltage at terminal A (V_A).

$V_A =$ _____

Questions: Does voltage V_A represent a binary "zero" or a binary "one"?

Is the logic probe light on or off? Does this indicate a binary "zero" or a binary "one"?

Why is the voltage less than 5 V?

Step 3.   Change resistor R to 2 kΩ. Click the On-Off switch to run the simulation again. Stop the simulation after the voltmeter reading is stable. Record the voltage at terminal A (V_A).

$V_A =$ _____

Questions: Was voltage V_A the same as in Step 2? **If not, why not?**

Does this represent a binary "zero" or a binary "one?" Does the logic probe verify this?

Step 4.   Change Resistor R to 500 Ω. Click the On-Off switch to run the simulation again. Stop the simulation after the voltmeter reading is stable. Record the voltage at terminal A (V_A).

$V_A = $ _____

Questions: Is the logic probe light still on? **If not, why not?**

Does voltage V_A represent a binary "zero" or a binary "one?" **Explain.**

Name_____

Date_____

EXPERIMENT

2 Logic Gates: INVERTER, OR, and AND

Objectives:

1. Investigate the behavior of the INVERTER logic gate.
2. Investigate the behavior of a two-input OR logic gate.
3. Investigate the behavior of a two-input AND logic gate.
4. Determine the effect of loading on a digital logic gate output.

Materials:

One 5 V dc voltage supply
Two logic switches
Three dc voltmeters
Three logic probe lights
One logic analyzer
Two pulse generators
One INVERTER (1-7404 IC)
One two-input OR gate (1-7432 IC)
One two-input AND gate (1-7408 IC)
Resistors—50 Ω, 200 Ω, 1 kΩ (2), 10 kΩ

Theory:

The **INVERTER logic gate** inverts the input. Therefore, when the input is low (0) the output will be high (1) and when the input is high (1) the output will be low (0). Circuits for studying the INVERTER logic gate are shown in Figures 2-1 and 2-2.

The **OR gate** output will be high (1) when one or more inputs are high (1). The OR gate output will be low (0) only when all of the inputs are low (0). Circuits for studying the two-input OR logic gate are shown in Figures 2-3 and 2-4.

The **AND gate** output will be high (1) only when all of the inputs are high (1). If any of the AND gate inputs are low (0), then the output will be low (0). Circuits for studying the two-input AND logic gate are shown in Figures 2-5 and 2-6.

If the **load resistance** on any of the TTL logic gates is too low and draws too much current from the gate output terminal, the output voltage will be **dragged down below 2 V** when the output is trying to produce a high (1) output. This will produce an **error** in the logical network output because other TTL logic gate inputs will not recognize an input below 2 V as a logical high (1). **Fan out** for any TTL logic gate is the number of TTL logic gate inputs that can be connected to a TTL logic gate output before the above problem of **overloading** will occur.

Figure 2-1 The INVERTER Gate

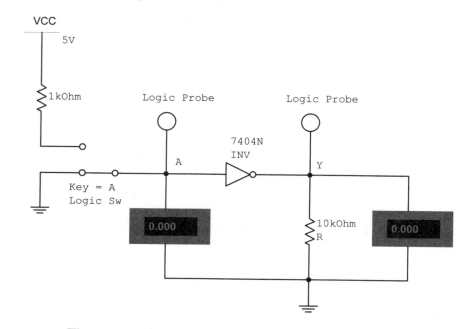

Figure 2-2 INVERTER Pulse Response (Multisim)

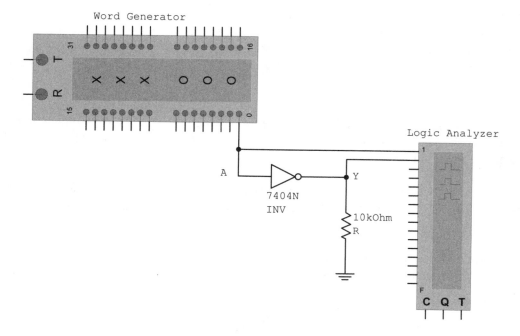

Figure 2-3 The OR Gate

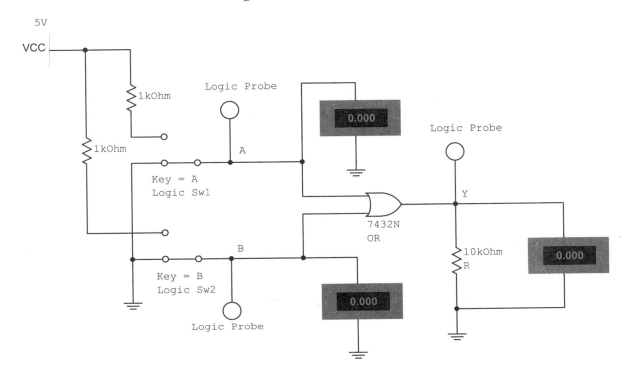

Figure 2-4 OR Gate Pulse Response (Multisim)

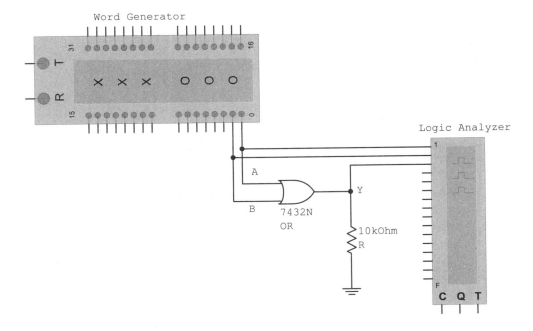

Figure 2-5 The AND Gate

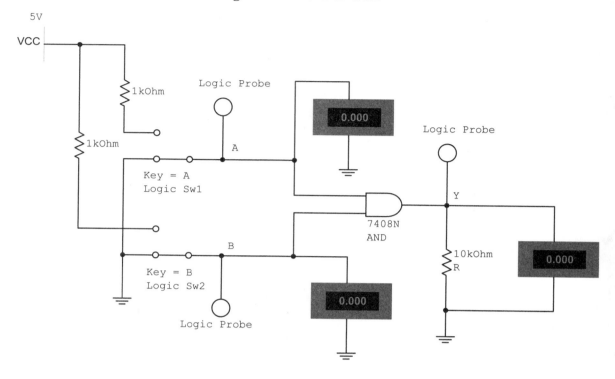

Figure 2-6 AND Gate Pulse Response (Multisim)

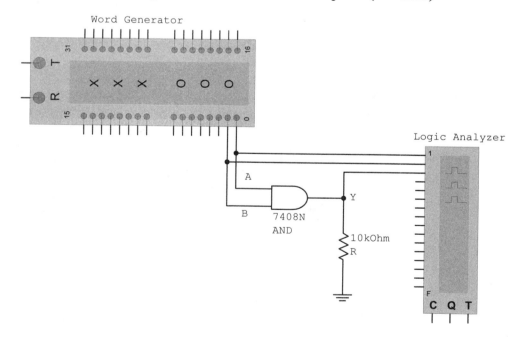

$$V = I \cdot R$$

Procedure:

INVERTER

Step 1.   Open circuit file FIG2-1. Notice that the logic switch is placing a zero (ground equals 0 V) on the input (A) of the INVERTER gate. Resistor R represents the load on the INVERTER output. Click the On-Off switch to run the simulation. Stop the simulation when the voltage readings are stable. Record the INVERTER input voltage (V_A) and output voltage (V_Y).

$V_A =$ ___0.011___ ✓ $V_Y =$ ___4.102 v___

Questions: Is the output probe light on or off? Does this represent a binary "zero" or a binary "one" on the INVERTER output?   OUT LIGHTS = ON
                                                                                          ONE

Is the input probe light on or off? Does this represent a binary "zero" or a binary "one" on the INVERTER input?   OFF = ZERO

Is the INVERTER output voltage above 2 V? Does this represent a binary "zero" or a binary "one?"
                                                        YES   = ONE

Why is the INVERTER output voltage less than 5 V?
                    BECAUSE IT GOES UP TO 5V
                                      FROM 2

Step 2.   Change Resistor R to 200 Ω. Click the On-Off switch to run the simulation again. Stop the simulation when the voltage readings are stable. Record the INVERTER output voltage (V_Y).

$V_Y =$ ___3.074 J___

Questions: How does the new INVERTER output voltage compare with the output voltage in Step 1? Why are they different?   SMALLER   LESS
                    LOWER, BECAUSE THE BIGGER RESISTOR DRAWS MORE VOLTAGE.

Step 3.   Change Resistor R to 50 Ω. Click the On-Off switch to run the simulation again. Stop the simulation when the voltage readings are stable. Record the INVERTER output voltage (V_Y).

$V_Y =$ ___1.074 v___

Questions: Is the INVERTER output voltage above 2 V?

No

Is the logic probe light on? **If not, why not?**

No, BECAUSE IT'S BELOW 2v

Step 4.   Change resistor R back to 10 kΩ. Press the "A" key on the keyboard to change the logic switch to a binary "one" input (up). Click the On-Off switch to run the simulation again. Stop the simulation when the voltage readings are stable. Record the INVERTER input voltage (V_A) and output voltage (V_Y).

$V_A =$ _4.673 v_ $V_Y =$ _0.110 v_

Questions: Is the output probe light on or off? Does this represent a binary "zero" or a binary "one" on the INVERTER output?   OFF = ZERO

Is the input probe light on or off? Does this represent a binary "zero" or a binary "one" on the INVERTER input?   ON = ONE

Is the INVERTER output voltage below 0.8 V? Does this represent a binary "zero" or a binary "one"?

YES, ZERO

What is the relationship between the binary input and the binary output for the INVERTER gate?

THEY'RE OPPOSITE

Step 5.   Open circuit file FIG2-2. Bring down the word generator enlargement and make sure that the following settings are selected: Frequency = 1 kHz, Trigger = Internal, Hex codes (00000000, 00000000, 00000001, 00000001, repeated four times). Move the word generator to the left by clicking and dragging. Bring down the logic analyzer enlargement and make sure that the following settings are selected: Clocks/Div = 16, Clock Setup (Clock Source = Internal, Clock Rate = 10 kHz, Pre-trigger Samples = 100, Post-trigger Samples = 1000, Threshold Voltage = 2.5), Trigger Settings (Trigger Clock Edge = Positive, Trigger Qualifier = x, Pattern A= xxxxxxxxxxxxxxxx, Trigger Combinations = A). Click BURST on the word generator to run the simulation. Notice that the word generator has applied a pulse pattern to the INVERTER input, shown in red on the logic analyzer screen. The blue curve plot is the INVERTER output. Draw the INVERTER input and output curve plots in the space provided and label them, then click the On-Off switch to end the simulation.

NOTE: In a hardwired laboratory, use a pulse generator to apply a square wave to the INVERTER input and use a dual trace oscilloscope to monitor the INVERTER input and output, if a logic analyzer is not available.

Questions: What is the relationship between the INVERTER input and output curve plots? Is it what you expected for an INVERTER?   *YES, THEY'RE OPPOSITES*

OR Gate

Step 6.   Open circuit file FIG2-3. Notice that Logic Switches 1 and 2 are placing binary "zeros" (ground equals 0 V) on the two OR gate inputs (A and B). Resistor R represents the load on the OR gate output. Click the On-Off switch to run the simulation. Record the output logic level (Y) for the binary inputs (A and B) in Table 2-1.

Question: Is the OR gate output voltage below 0.8 V when a binary "zero" is applied to both gate inputs? Is this output a binary "one" or a binary "zero"? Do the logic probe lights confirm this?

          YES, ZERO, YES

Table 2-1 OR Gate

A	B	Y
0	0	*0*
0	1	*1*
1	0	*1*
1	1	*1*

Step 7.   By pressing the "A" and "B" keys on the keyboard to change the logic switch positions, change the binary inputs (A and B) to the remaining values in Table 2-1 and record the output logic level (Y) for each case, then stop the simulation.

Questions: Is the OR gate output voltage above 2 V when a binary "one" is applied to one or more of the inputs? Is this output a binary "one" or a binary "zero"? Do the logic probe lights confirm this?

          YES, ONE, YES

Based on the results in Table 2-1, what conclusion can you draw about the relationship between the OR gate binary output and the binary inputs?

ANY INPUT = 1, OUTPUT = 1

Step 8.   Open circuit file FIG2-4. Bring down the word generator enlargement and make sure that the following settings are selected: Frequency = 1 kHz, Trigger = Internal, Hex codes (00000000, 00000001, 00000002, 00000003, repeated four times). Move the word generator to the left by clicking and dragging. Bring down the logic analyzer enlargement. The logic analyzer settings are the same as in Step 5. Click BURST on the word generator to run the simulation. Notice that the word generator has applied a pulse pattern to each OR gate input, shown in red and green on the logic analyzer screen. The blue curve plot is the OR gate output. Draw the OR gate input and output curve plots in the space provided and label them, then stop the simulation.

NOTE: In a hardwired laboratory, use pulse generator outputs to apply square waves to the OR gate inputs.

Question: Does the OR gate output go low only when both inputs are low? Is this expected for an OR gate?

YES, YES

AND Gate

Step 9. Open circuit file FIG2-5. Notice that Logic Switches 1 and 2 are placing binary zeros (ground equals 0 V) on the two AND gate inputs (A and B). Resistor R represents the load on the AND gate output. Click the On-Off switch to run the simulation. Record the output logic level (Y) for the binary inputs (A and B) in Table 2-2.

Question: Is the AND gate output voltage below 0.8 V when a binary "zero" is applied to one or more of the inputs? Is this output a binary "one" or a binary "zero"? Do the logic probe lights confirm this?

YES ,   ZERO   ,   YES

Table 2-2 AND Gate

A	B	Y
0	0	0
0	1	0
1	0	0
1	1	1

Step 10. By pressing the "A" and "B" keys on the keyboard to change the logic switch positions, change the binary inputs (A and B) to the remaining values in Table 2-2 and record the output logic level (Y) for each case, then stop the simulation.

Questions: Is the AND gate output voltage above 2 V when a binary "one" is applied to both gate inputs? Is this output a binary "one" or a binary "zero"? Do the logic probe lights confirm this?

YES,   ONE ,   YES

Based on the results in Table 2-2, what conclusion can you draw about the relationship between the AND gate binary output and the binary inputs?

THE OUTPUT WILL BE 1 ONLY WHEN
ALL INPUTS ARE 1

Step 11.   Open circuit file FIG2-6. The word generator and logic analyzer settings should be the
            same as in Step 8. Click BURST on the word generator to run the simulation. Notice that
            the word generator has applied a pulse pattern to each AND gate input, shown in red and
            green on the logic analyzer screen. The blue curve plot is the AND gate output. Draw the
            AND gate input and output curve plots in the space provided and label them, then stop the
            simulation.

NOTE: In a hardwired laboratory, use pulse generator outputs to apply square waves to the AND
gate inputs.

Question: Does the AND gate output go high only when both inputs are high? Is this expected for an
AND gate?

YES,   YES

EXPERIMENT

Logic Gates: NAND and NOR

Objectives:

1. Complete the truth table for an AND gate.
2. Complete the truth table for a NAND gate and compare it with the AND gate truth table.
3. Plot the NAND gate pulse response timing diagram.
4. Complete the truth table for an OR gate.
5. Complete the truth table for a NOR gate and compare it with the OR gate truth table.
6. Plot the NOR gate pulse response timing diagram.

Materials:

One 5 V dc voltage supply
Two logic switches
Three logic probe lights
One logic analyzer
Two pulse generators
One two-input OR gate (1-7432 IC)
One two-input AND gate (1-7408 IC)
One two-input NAND gate (1-7400 IC)
One two-input NOR gate (1-7402 IC)
Resistors—1 kΩ (2) and 10 kΩ

Theory:

A **truth table** lists the outputs for all of the possible input combinations for a logic gate or logic network. See the Theory section in Experiment 2 for information on the OR and AND logic gates.

The **NAND gate** is equivalent to an AND gate with an INVERTER on the output. Therefore, the NAND gate output is the **inverse** of the **AND gate output**. The NAND gate output will be low (0) only when all of the inputs are high (1). The NAND gate output will be high (1) if any of the inputs are low (0). Circuits for plotting the truth table for the two-input AND and NAND logic gates are shown in Figures 3-1 and 3-2, respectively. The circuit for plotting a NAND gate **pulse response timing diagram** is shown in Figure 3-3.

The **NOR gate** is equivalent to an OR gate with an INVERTER on the output. Therefore, the NOR gate output is the **inverse** of the **OR gate output**. The NOR gate output will be high (1) only when all of the inputs are low (0). The NOR gate output will be low (0) if any of the inputs are high (1). Circuits for plotting the truth table for the two-input OR and NOR logic gates are shown in Figures 3-4 and 3-5 respectively. The circuit for plotting a NOR gate **pulse response timing diagram** is shown in Figure 3-6.

Figure 3-1 AND Gate

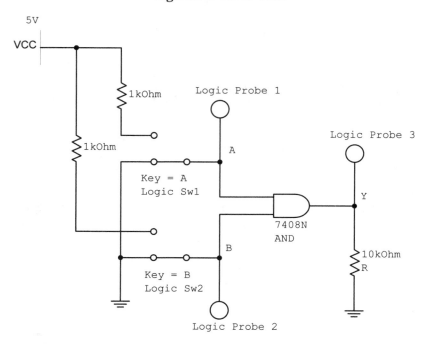

Figure 3-2 NAND Gate

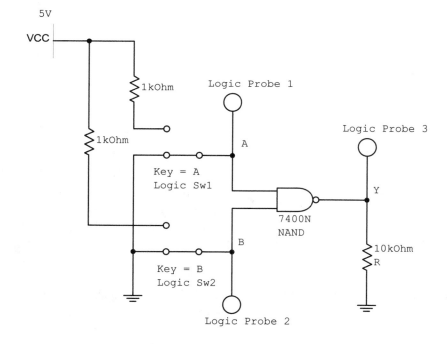

Figure 3-3 NAND Gate Pulse Response (Multisim)

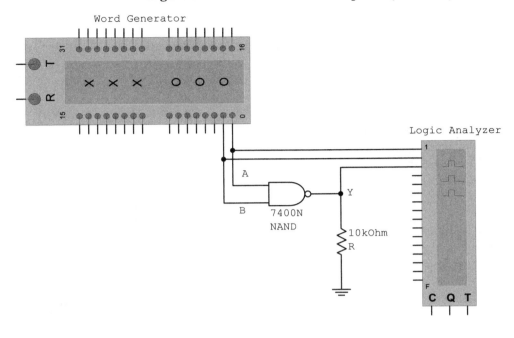

Figure 3-4 OR Gate

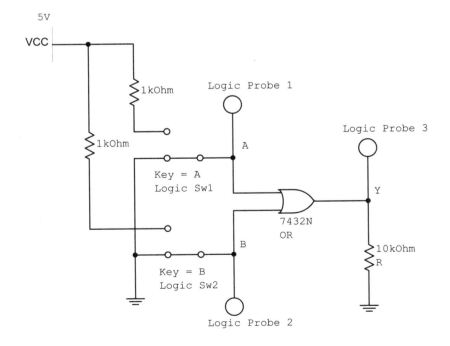

Figure 3-5 NOR Gate

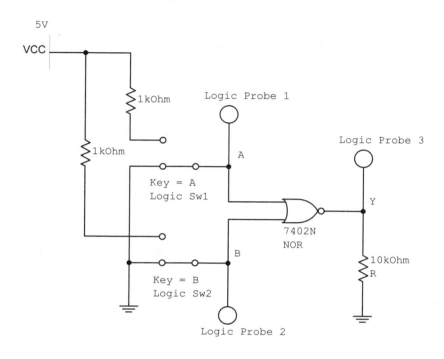

Figure 3-6 NOR Gate Pulse Response (Multisim)

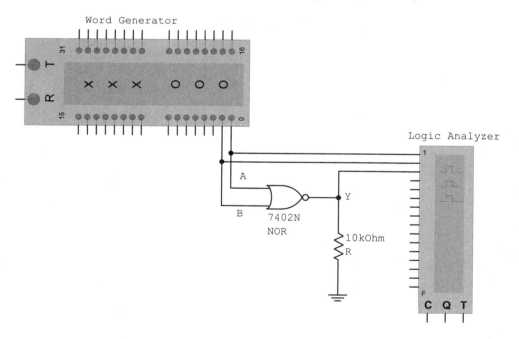

Procedure:

Step 1.   Open circuit file FIG3-1. Notice that Logic Switches 1 and 2 are placing binary "zeros" (ground equals 0 V) on the two AND gate inputs (A and B). Resistor R represents the load on the AND gate output. Click the On-Off switch to run the simulation. By switching the logic switches to the appropriate positions, complete the AND gate truth table (Table 3-1). To switch Logic Switch 1, type "A" on the keyboard. To switch Logic Switch 2, type "B" on the keyboard. After the truth table is complete, stop the simulation.

Table 3-1 AND Gate Truth Table

A	B	Y
0	0	0
0	1	0
1	0	0
1	1	1

Step 2.   Open circuit file FIG3-2. Click the On-Off switch to run the simulation. By switching the logic switches to the appropriate positions, complete the NAND gate truth table (Table 3-2). After the truth table is complete, stop the simulation.

Table 3-2 NAND Gate Truth Table

A	B	Y
0	0	1
0	1	1
1	0	1
1	1	0

Question: What is the difference between the AND gate truth table and the NAND gate truth table? **Explain.**

THEY ARE OPPOSITES. THE NAND GATE HAS AN INVERTER AT THE END

Step 3.   Open circuit file FIG3-3. Bring down the word generator enlargement and make sure that the following settings are selected: Frequency = 1 kHz, Trigger = Internal, Hex codes (00000000, 00000001, 00000002, 00000003, repeated four times). Move the word generator to the left by clicking and dragging. Bring down the logic analyzer enlargement and make sure that the following settings are selected: Clocks/Div = 16, Clock Setup (Clock Source = Internal, Clock Rate = 10 kHz, Pre-trigger Samples = 100, Post-trigger Samples = 1000, Threshold Voltage = 2.5), Trigger Settings (Trigger Clock Edge = Positive, Trigger Qualifier = x, Pattern A= xxxxxxxxxxxxxxxx, Trigger Combinations = A). Click BURST on the word generator to run the simulation. Notice that the word generator has applied a pulse pattern to each NAND gate input, shown in red and green on the logic analyzer screen. The blue curve plot is the NAND gate output. Draw the NAND gate input and output curve plots in the space provided and label them, then stop the simulation.

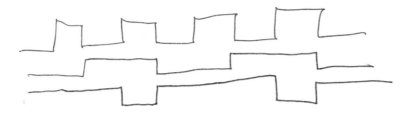

NOTE: In a hardwired laboratory, use pulse generator outputs to apply square waves to the NAND gate inputs.

Question: Does the NAND gate output go low only when all of the inputs are high? Is this expected for a NAND gate?

YES,   YES

Step 4.   Open circuit file FIG3-4. Notice that Logic Switches 1 and 2 are placing binary zeros (ground equals 0 V) on the two OR gate inputs (A and B). Resistor R represents the load on the OR gate output. Click the On-Off switch to run the simulation. By switching the logic switches to the appropriate positions, complete the OR gate truth table (Table 3-3). After the truth table is complete, stop the simulation.

Table 3-3 OR Gate Truth Table

A	B	Y
0	0	0
0	1	1
1	0	1
1	1	1

Step 5.   Open circuit file FIG3-5. Click the On-Off switch to run the simulation. By switching the logic switches to the appropriate positions, complete the NOR gate truth table (Table 3-4). After the truth table is complete, stop the simulation.

Table 3-4 NOR Gate Truth Table

A	B	Y
0	0	1
0	1	0
1	0	0
1	1	0

Question: What is the difference between the OR gate truth table and the NOR gate truth table?
Explain.

THEY'RE opposites, NoR GATE HAS AN inVERTER in THE oUTPUT

Step 6.   Open circuit file FIG3-6. The word generator and logic analyzer settings should be the same as in Step 3. Click BURST on the word generator to run the simulation. Notice that the word generator has applied a pulse pattern to each NOR gate input, shown in red and green on the logic analyzer screen. The blue curve plot is the NOR gate output. Draw the NOR gate input and output curve plots in the space provided and label them, then stop the simulation.

NOTE: In a hardwired laboratory, use pulse generator outputs to apply square waves to the NOR gate inputs.

Question: Does the NOR gate output go high only when all of the inputs are low? Is this expected for a NOR gate?

YES , yes

EXPERIMENT

4 Boolean Theorems— DeMorgan's Theorems

Objectives:

1.   Verify experimentally some of the multivariable Boolean theorems.
2.   Verify experimentally DeMorgan's theorems.

Materials:

One 5 V dc voltage supply
Three logic switches
Four logic probe lights
Two two-input AND gates (1-7408 IC)
Two INVERTERS (1-7404 IC)
One two-input OR gate (1-7432 IC)
Three 1 kΩ resistors

Theory:

The operation of a **logic circuit** consisting of a **combination of logic gates** can be described with a **Boolean equation**. The Boolean equation for the output of a logic circuit is obtained by first writing the Boolean equation for the output of each **logic gate**, working from the logic circuit input towards the logic circuit output. In the resulting Boolean equation, all AND operations are assumed to be performed before the OR operations unless an OR operation is surrounded by parentheses, in which case the OR operation is performed first.

The **Boolean equation** for a logic circuit can be simplified by using one or more of the **Boolean theorems**. A simplified logic circuit can then be drawn from the simplified Boolean equation. In this experiment, you will verify some of the Boolean theorems that you will be using to simplify logic circuits.

The single-variable Boolean theorems involving the AND operation are

$$(X)(0) = 0$$
$$(X)(1) = X$$
$$(X)(X) = X$$
$$(X)(X') = 0$$

The single-variable Boolean theorems involving the OR operation are

$$X + 0 = X$$
$$X + 1 = 1$$
$$X + X = X$$
$$X + X' = 1$$

Figure 4-1, 4-2, 4-3, and 4-4 show circuits that will be used to verify the following multivariable Boolean theorems.

$$X(Y + Z) = XY + XZ$$
$$X + XY = X$$
$$X + X'Y = X + Y$$

Figures 4-5 and 4-6 show circuits that will be used to verify the following **DeMorgan's theorems**.

$$(XY)' = X' + Y'$$
$$(X + Y)' = X'Y'$$

DeMorgan's theorems show that the NAND gate is equivalent to an OR gate with INVERTERS on the inputs, and the NOR gate is equivalent to an AND gate with INVERTERS on the inputs.

Figure 4-1 Multivariable Boolean Theorem

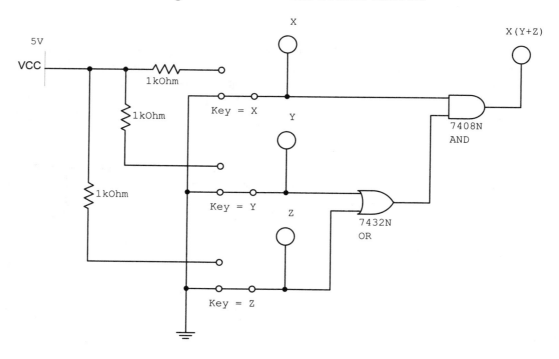

Figure 4-2 Multivariable Boolean Theorem

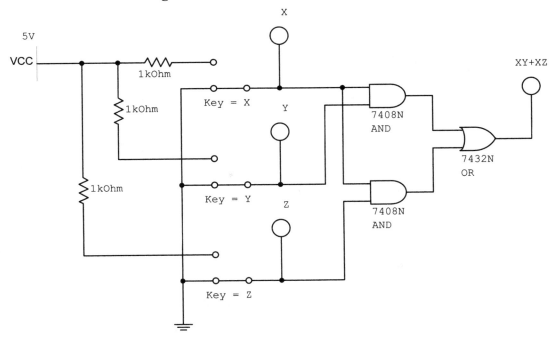

Figure 4-3 Multivariable Boolean Theorem

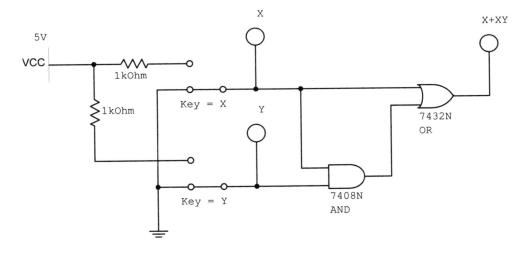

Figure 4-4 Multivariable Boolean Theorem

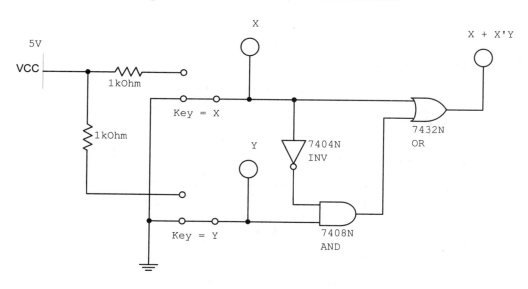

Figure 4-5 DeMorgan's Theorem

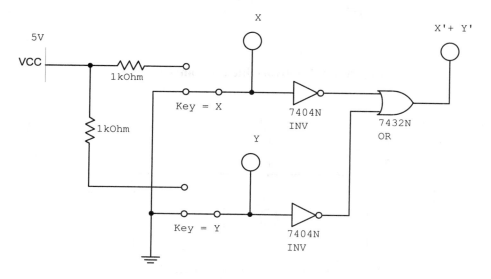

Figure 4-6 DeMorgan's Theorem

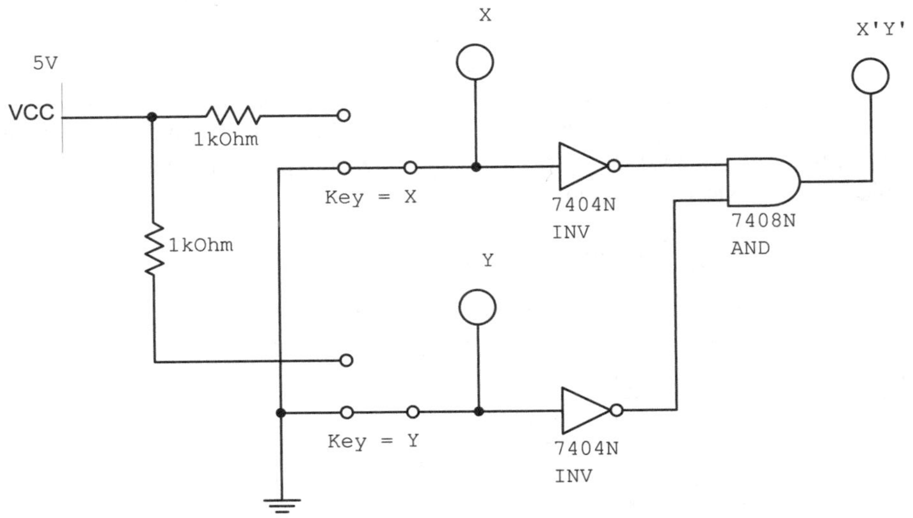

Procedure:

Step 1.   Open circuit file FIG4-1. Click the On-Off switch to run the simulation. Press the "X", "Y", and "Z" keys on the keyboard to switch inputs X, Y, and Z between a low (0) and a high (1) binary input to the logic network. Record the output X(Y + Z) for each input combination on the truth table (Table 4-1), then stop the simulation.

Table 4-1 Truth Table

X	Y	Z	X(Y + Z)	XY + XZ
0	0	0		
0	0	1		
0	1	0		
0	1	1		
1	0	0		
1	0	1		
1	1	0		
1	1	1		

Step 2.   Open circuit file FIG4-2. Click the On-Off switch to run the simulation. Press the "X", "Y", and "Z" keys on the keyboard to switch inputs X, Y, and Z between a low (0) and a high (1) binary input to the logic network. Record the output XY + XZ for each input combination on the truth table (Table 4-1). Then stop the simulation.

Question: Do these results correspond with the known Boolean theorem $X(Y + Z) = XY + XZ$?

Step 3.   Open circuit file FIG4-3. Click the On-Off switch to run the simulation. Press the "X" and "Y" keys on the keyboard to switch inputs X and Y between a low (0) and a high (1) binary input to the logic network. Record the output $X + XY$ for each input combination on the truth table (Table 4-2), then stop the simulation.

Table 4-2 Truth Table

X	Y	X + XY
0	0	
0	1	
1	0	
1	1	

Question: Do these results correspond with the known Boolean theorem $X = X + XY$?

Step 4.   Open circuit file FIG4-4. Click the On-Off switch to run the simulation. Press the "X" and "Y" keys on the keyboard to switch inputs X and Y between a low (0) and a high (1) binary input to the logic network. Record the output $X + X'Y$ for each input combination on the truth table (Table 4-3), then stop the simulation.

Table 4-3 Truth Table

X	Y	X + X'Y	X + Y
0	0		
0	1		
1	0		
1	1		

Step 5.   Record the expected outputs for an OR gate in the $X + Y$ column in Table 4-3. (See Experiment 3, Step 4.)

Question: Do these results correspond with the known Boolean theorem $X + X'Y = X + Y$?

Step 6.   Open circuit file FIG4-5. Click the On-Off switch to run the simulation. Press the "X" and "Y" keys on the keyboard to switch inputs X and Y between a low (0) and a high (1) binary input to the logic network. Record the output $X' + Y'$ for each input combination on the truth table (Table 4-4), then stop the simulation.

Table 4-4 Truth Table

X	Y	$X' + Y'$	$(XY)'$
0	0		
0	1		
1	0		
1	1		

Step 7.   Record the expected outputs for a NAND gate in the $(XY)'$ column in Table 4-4. (See Experiment 3, Step 2.)

Question: Do these results correspond with DeMorgan's theorem $(XY)' = X' + Y'$?

Step 8.   Open circuit file FIG4-6. Click the On-Off switch to run the simulation. Press the "X" and "Y" keys on the keyboard to switch inputs X and Y between a low (0) and a high (1) binary input to the logic network. Record the output $X'Y'$ for each input combination on the truth table (Table 4-5), then stop the simulation.

Table 4-5 Truth Table

X	Y	$X'Y'$	$(X + Y)'$
0	0		
0	1		
1	0		
1	1		

Step 9.   Record the expected outputs for a NOR gate in the $(X + Y)'$ column in Table 4-5. (See Experiment 3, Step 5.)

Question: Do these results correspond with DeMorgan's theorem $(X + Y)' = X'Y'$?

EXPERIMENT

Universality of NAND and NOR Gates

Objectives:

1. Show how an INVERTER can be implemented using a NAND gate.
2. Show how an AND gate can be implemented using NAND gates.
3. Show how an OR gate can be implemented using NAND gates.
4. Show how a NOR gate can be implemented using NAND gates.
5. Show how an INVERTER can be implemented using a NOR gate.
6. Show how an OR gate can be implemented using NOR gates.
7. Show how an AND gate can be implemented using NOR gates.
8. Show how a NAND gate can be implemented using NOR gates.

Materials:

One 5 V dc voltage supply
Two logic switches
Three logic probe lights
Four two-input NAND gates (1-7400 IC)
Four two-input NOR gates (1-7402 IC)
Resistors: 1 kΩ (2) and 10 kΩ

Theory:

See the Theory section of Experiment 2 for information on INVERTERS, OR gates, and AND gates.
See the Theory section of Experiment 3 for information on truth tables, NAND gates, and NOR gates.
See the Theory section of Experiment 4 for information on the Boolean theorems and DeMorgan's theorems.

When the inputs of a NAND gate are connected together, as shown in Figure 5-1, the NAND gate will act like an INVERTER because

$X = (AA)' = A'$.

When a NAND gate INVERTER is connected to the output of a NAND gate, as shown in Figure 5-2, the logic circuit will act like an AND gate because

$X = (AB)'' = AB$.

When two NAND gate INVERTERS are connected to the inputs of a NAND gate, as shown in Figure 5-3, the logic circuit will act like an OR gate because

$X = (A'B')'$, and $A'B' = (A + B)'$ from DeMorgan's theorem.

Therefore, $X = (A + B)'' = A + B$.

In the logic circuit in Figure 5-4, a NAND gate INVERTER is connected to the output of the OR logic circuit in Figure 5-3, causing it to act like a NOR gate because

$X = (A + B)'$.

When the inputs of a NOR gate are connected together, as shown in Figure 5-5, the NOR gate will act like an INVERTER because

$X = (A + A)' = A'$.

When a NOR gate INVERTER is connected to the output of a NOR gate, as shown in Figure 5-6, the logic circuit will act like an OR gate because

$X = (A + B)'' = A + B$.

When two NOR gate INVERTERS are connected to the inputs of a NOR gate, as shown in Figure 5-7, the logic circuit will act like an AND gate because

$X = (A' + B')'$, and $A' + B' = (AB)'$ from DeMorgan's theorem.

Therefore, $X = (AB)'' = AB$.

In the logic circuit in Figure 5-8, a NOR gate INVERTER is connected to the output of the AND logic circuit in Figure 5-7, causing it to act like a NAND gate because

$X = (AB)'$.

Figure 5-1 NAND Gate INVERTER

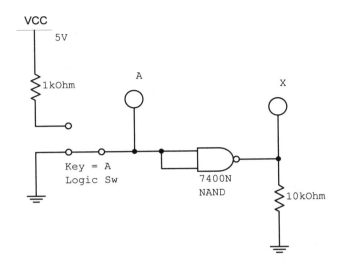

Figure 5-2 NAND Gate Implementation of an AND Gate

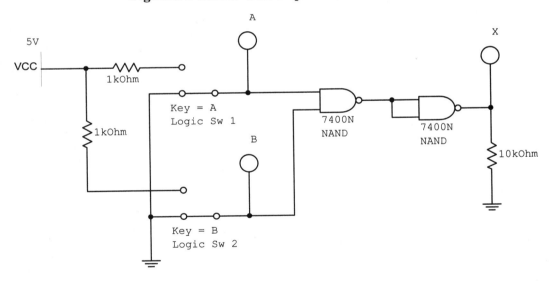

Figure 5-3 NAND Gate Implementation of an OR Gate

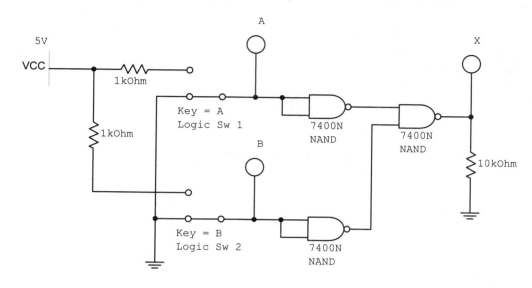

Figure 5-4 NAND Gate Implementation of a NOR Gate

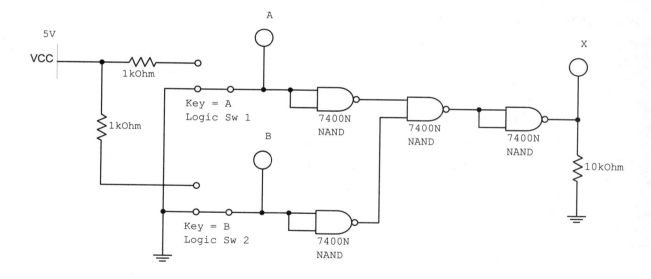

Figure 5-5 NOR Gate INVERTER

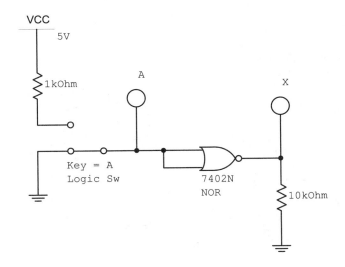

Figure 5-6 NOR Gate Implementation of an OR Gate

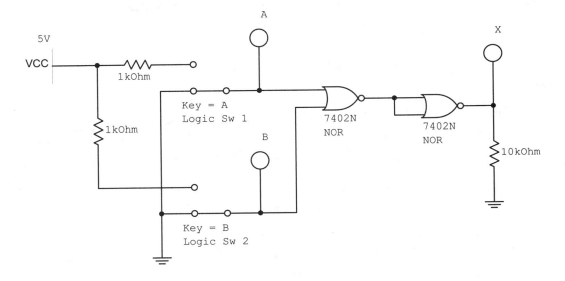

Figure 5-7 NOR Gate Implementation of an AND Gate

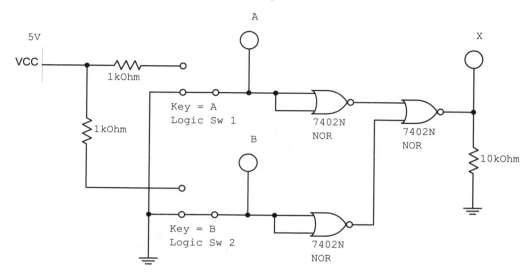

Figure 5-8 NOR Gate Implementation of a NAND Gate

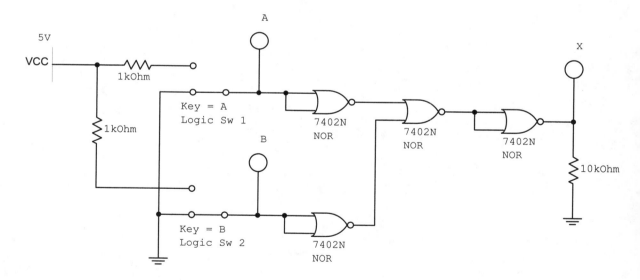

Procedure:

Step 1.   Open circuit file FIG5-1. Determine the Boolean equation for the logic circuit in Figure 5-1.

Step 2.   Click the On-Off switch to run the simulation. By switching the logic switch to the appropriate positions, complete the truth table (Table 5-1), then stop the simulation. The logic switch can be switched by pressing the "A" key on the keyboard.

Table 5-1 Truth Table

A	X
0	
1	

Question: Based on the logic equation in Step 1 and the truth table in Step 2, what is the logic circuit in Figure 5-1 equivalent to?

Step 3.   Open circuit file FIG5-2. Determine the Boolean equation for the logic circuit in Figure 5-2.

Step 4.   Click the On-Off switch to run the simulation. By switching the logic switches to the appropriate positions, complete the truth table (Table 5-2), then stop the simulation. The logic switches can be switched by pressing the "A" and "B" keys on the keyboard.

Table 5-2 Truth Table

A	B	X
0	0	
0	1	
1	0	
1	1	

Question: Based on the logic equation in Step 3 and the truth table in Step 4, what is the logic circuit in Figure 5-2 equivalent to?

Step 5.   Open circuit file FIG5-3. Determine the Boolean equation for the logic circuit in Figure 5-3.

Step 6.   Click the On-Off switch to run the simulation. By switching the logic switches to the appropriate positions, complete the truth table (Table 5-3), then stop the simulation. The logic switches can be switched by pressing the "A" and "B" keys on the keyboard.

Table 5-3 Truth Table

A	B	X
0	0	
0	1	
1	0	
1	1	

Question: Based on the logic equation in Step 5 and the truth table in Step 6, what is the logic circuit in Figure 5-3 equivalent to?

Step 7.   Open circuit file FIG5-4. Determine the Boolean equation for the logic circuit in Figure 5-4.

Step 8.   Click the On-Off switch to run the simulation. By switching the logic switches to the appropriate positions, complete the truth table (Table 5-4), then stop the simulation. The logic switches can be switched by pressing the "A" and "B" keys on the keyboard.

Table 5-4 Truth Table

A	B	X
0	0	
0	1	
1	0	
1	1	

Question: Based on the logic equation in Step 7 and the truth table in Step 8, what is the logic circuit in Figure 5-4 equivalent to?

Step 9.   Open circuit file FIG5-5. Determine the Boolean equation for the logic circuit in Figure 5-5.

Step 10.   Click the On-Off switch to run the simulation. By switching the logic switch to the appropriate positions, complete the truth table (Table 5-5), then stop the simulation. The logic switch can be switched by pressing the "A" key on the keyboard.

Table 5-5 Truth Table

A	X
0	
1	

Question: Based on the logic equation in Step 9 and the truth table in Step 10, what is the logic circuit in Figure 5-5 equivalent to?

Step 11.   Open circuit file FIG5-6. Determine the Boolean equation for the logic circuit in Figure 5-6.

Step 12.   Click the On-Off switch to run the simulation. By switching the logic switches to the
            appropriate positions, complete the truth table (Table 5-6), then stop the simulation. The
            logic switches can be switched by pressing the "A" and "B" keys on the keyboard.

Table 5-6 Truth Table

A	B	X
0	0	
0	1	
1	0	
1	1	

Question: Based on the logic equation in Step 11 and the truth table in Step 12, what is the logic
circuit in Figure 5-6 equivalent to?

Step 13.   Open circuit file FIG5-7. Determine the Boolean equation for the logic circuit in Figure 5-7.

Step 14.   Click the On-Off switch to run the simulation. By switching the logic switches to the
            appropriate positions, complete the truth table (Table 5-7), then stop the simulation. The
            logic switches can be switched by pressing the "A" and "B" keys on the keyboard.

Table 5-7 Truth Table

A	B	X
0	0	
0	1	
1	0	
1	1	

Question: Based on the logic equation in Step 13 and the truth table in Step 14, what is the logic
circuit in Figure 5-7 equivalent to?

Step 15.   Open circuit file FIG5-8. Determine the Boolean equation for the logic circuit in Figure 5-8.

Step 16.   Click the On-Off switch to run the simulation. By switching the logic switches to the
            appropriate positions, complete the truth table (Table 5-8), then stop the simulation. The
            logic switches can be switched by pressing the "A" and "B" keys on the keyboard.

Table 5-8 Truth Table

A	B	X
0	0	
0	1	
1	0	
1	1	

Question: Based on the logic equation in Step 15 and the truth table in Step 16, what is the logic
circuit in Figure 5-8 equivalent to?

Name_____

Date_____

EXPERIMENT

Analyzing Combinational Logic Circuits

Objectives:

1. Learn how to determine the Boolean equation for a logic circuit.
2. Learn how to derive the truth table for a logic circuit.
3. Learn how to convert AND-OR logic to NAND-NAND logic.
4. Learn how to convert OR-AND logic to AND-OR logic.
5. Learn how to convert NAND-NAND logic to AND-OR logic.

Materials:

One 5 V dc voltage supply
One logic probe light
Three logic switches
Two two-input AND gates (1-7408 IC)
One two-input OR gate (1-7432 IC)
Three two-input NAND gates (1-7400 IC)
One INVERTER (1-7404 IC)
Three 1 kΩ resistors

Theory:

A **logic circuit** consisting of a **combination of logic gates** is called a **combinational logic circuit** and can be described with a **Boolean equation**. The Boolean equation for the output of a logic circuit is obtained by first writing the Boolean equation for the output of each logic gate, working from the logic circuit input towards the logic circuit output. In the resulting Boolean equation, all AND operations are assumed to be performed before the OR operations unless an OR operation is surrounded by parentheses, in which case the OR operation is performed first.

A **truth table** for a logic circuit shows the outputs for all possible input combinations. The truth table can be developed from the Boolean equation by evaluating the equation outputs for all possible input combinations. This is accomplished by first evaluating each part of the equation for all possible input combinations, taking operations surrounded by parentheses first, AND operations next, and OR operations last. If an equation is inverted, perform the operation of the equation first, and then invert the result.

An **AND-OR logic circuit**, often called **sum-of-product** logic, consists of a series of AND gates feeding an OR gate. An **OR-AND logic circuit**, often called **product-of-sum** logic, consists of a series of OR gates feeding an AND gate. A **NAND-NAND logic circuit** consists of a series of NAND gates feeding a NAND gate.

An AND-OR logic circuit can be easily converted to NAND-NAND logic by adding INVERTERS to the outputs of all the AND gates and all the inputs of the OR gate. This results in double inversion between each AND gate output and the OR gate inputs, resulting in no change to the circuit operation because double inversion is the same as no inversion. Because the OR gate with inversion at the inputs is equivalent to a NAND gate (DeMorgan's theorem), the resulting equivalent circuit consists of all NAND gates.

A NAND-NAND logic circuit can be converted to AND-OR logic by representing the last NAND gate as an OR gate with inversion on the inputs (DeMorgan's theorem). This will result in double inversion between the outputs of the NAND gates and the OR gate inputs. Remove the double INVERTERS and an AND-OR logic circuit will result.

Review the results of Steps 1 and 2 in Experiment 4 to determine how to convert a product-of-sum (OR-AND) Boolean equation to a sum-of-products (AND-OR) Boolean equation. Review DeMorgan's theorems in the Theory section of Experiment 4 to determine how to simplify a logic equation for a NAND-NAND logic circuit.

The circuits in Figure 6-1, 6-2, 6-3, and 6-4 will be used to demonstrate how to determine the Boolean equation for a logic circuit, how to derive a truth table, how to convert an AND-OR logic circuit to NAND-NAND logic, how to convert an OR-AND logic circuit to AND-OR logic, and how to convert a NAND-NAND logic circuit to AND-OR logic:

Figure 6-1 AND-OR Combinational Logic Circuit

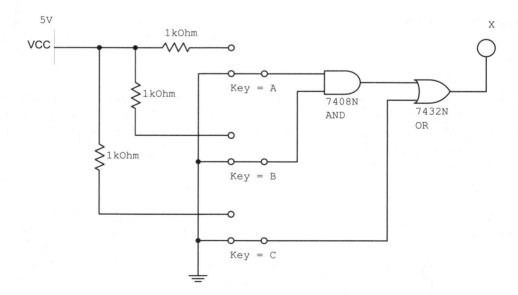

Figure 6-2 OR-AND Combinational Logic Circuit

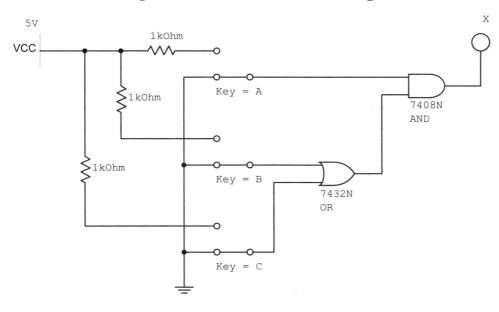

Figure 6-3 NAND-NAND Combinational Logic Circuit

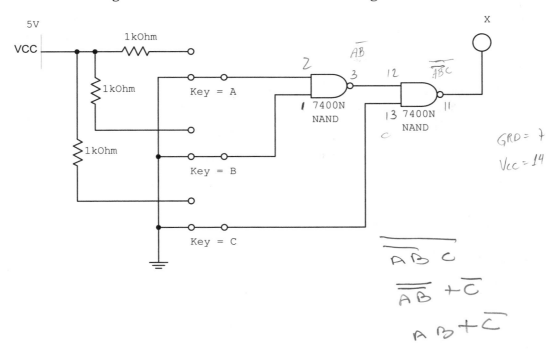

Figure 6-4 AND-OR Combinational Logic Circuit

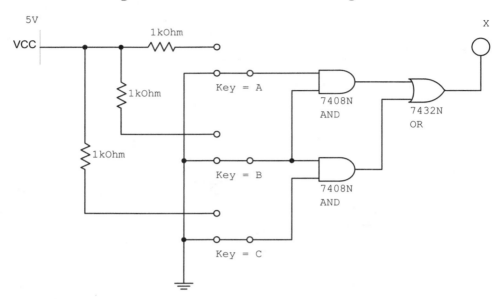

Procedure:

Step 1.   Open circuit file FIG6-1. Write the Boolean equation for the logic circuit in Figure 6-1.

Step 2.   Complete the truth table (Table 6-1) for the logic circuit in Figure 6-1 from the Boolean equation in Step 1. Use the space to the right of the truth table to evaluate each part of the Boolean equation.

Table 6-1 Truth Table

A	B	C	X		
0	0	0			
0	0	1			
0	1	0			
0	1	1			
1	0	0			
1	0	1			
1	1	0			
1	1	1			

Step 3.   Click the On-Off switch to run the simulation. By switching logic switches A, B, and C to
            the values of A, B, and C in Table 6-1 and monitoring the logic probe light (X), verify the
            truth table, then stop the simulation.

Question: Was the truth table verified by the logic circuit measurements?

Step 4.   Convert the AND-OR logic circuit shown in Figure 6-1 to NAND-NAND logic and draw
            the NAND-NAND equivalent circuit in the space provided.

Step 5.   Replace the AND-OR logic circuit in Figure 6-1 with the equivalent NAND-NAND logic
            circuit in Step 4. Click the On-Off switch to run the simulation. By switching logic
            switches A, B, and C to the values of A, B, and C in Table 6-1 and monitoring the logic
            probe light (X), verify the truth table, then stop the simulation.

Questions: Was the NAND-NAND truth table the same as the AND-OR truth table (Table 6-1)? Was
the NAND-NAND logic circuit equivalent to the AND-OR logic circuit?

Step 6.   Open circuit file FIG 6-2. Write the Boolean equation for the logic circuit in Figure 6-2.

Question: How does this Boolean equation compare with the Boolean equation for the circuit in
Figure 6-1 derived in Step 1? **Explain.**

Step 7.   Convert the Boolean equation in Step 6 to sum-of-products (AND-OR) form.

Step 8.   Complete the truth table (Table 6-2) for the logic circuit in Figure 6-2 from the Boolean equation in Step 7. Use the space to the right of the truth table to evaluate each part of the Boolean equation.

Table 6-2 Truth Table

A	B	C	X	A̅B̅	C̅
0	0	0	1	2	1
0	0	1		0	
0	1	0	1	0	1
0	1	1	0	0	
1	0	0	1	0	1
1	0	1	0	0	
1	1	0	1	1	1
1	1	1	1	1	

Step 9.   Click the On-Off switch to run the simulation. By switching logic switches A, B, and C to the values of A, B, and C in Table 6-2 and monitoring the logic probe light (X), verify the truth table, then stop the simulation.

Question: Was the truth table verified by the logic circuit measurements?

YES

Step 10.   Convert the logic circuit in Figure 6-2 to an AND-OR logic circuit from the sum-of-products equation derived in Step 7 and draw the AND-OR logic equivalent circuit in the space provided.

Step 11.   Replace the OR-AND logic circuit in Figure 6-2 with the equivalent AND-OR logic circuit
            in Step 10. Click the On-Off switch to run the simulation. By switching logic switches A,
            B, and C to the values of A, B, and C in Table 6-2 and monitoring the logic probe light (X),
            verify the truth table, then stop the simulation.

Questions: Was the AND-OR truth table the same as the OR-AND truth table (Table 6-2)? Was the
AND-OR logic circuit equivalent to the OR-AND logic circuit?

Step 12.   Convert the AND-OR logic circuit in Step 10 to NAND-NAND logic and draw the circuit in
            the space provided.

Step 13.   Replace the AND-OR logic circuit in Figure 6-2 with the equivalent NAND-NAND logic
            circuit in Step 12. Click the On-Off switch to run the simulation. By switching logic
            switches A, B, and C to the values of A, B, and C in Table 6-2 and monitoring the logic
            probe light (X), verify the truth table, then stop the simulation.

Questions: Was the NAND-NAND truth table the same as the AND-OR truth table (Table 6-2)? Was
the NAND-NAND logic circuit equivalent to the AND-OR logic circuit?

Step 14.   Open circuit file FIG6-3. Write the Boolean equation for the logic circuit in Figure 6-3.

$$X = \overline{\overline{ABC}} = \overline{\overline{AB} + C} = AB + \overline{C}$$

Step 15.   Simplify the logic equation in Step 14 into AND-OR (sum-of-products) logic using
            DeMorgan's theorem.

$$X = AB + \overline{C}$$

Step 16.   Complete the truth table (Table 6-3) for the logic circuit in Figure 6-3 using the simplified AND-OR Boolean equation derived in Step 15. Use the space to the right of the truth table to evaluate each part of the Boolean equation.

Table 6-3 Truth Table

A	B	C	X	AB	$\bar{C}$
0	0	0	1	0	1
0	0	1	0	0	0
0	1	0	1	0	1
0	1	1	0	0	0
1	0	0	1	0	1
1	0	1	0	0	0
1	1	0	1	1	1
1	1	1	1	1	0

Step 17.   Click the On-Off switch to run the simulation. By switching logic switches A, B, and C to the values of A, B, and C in Table 6-3 and monitoring the logic probe light (X), verify the truth table, then stop the simulation.

Question: Was the truth table verified by the logic circuit measurements?

Step 18.   Convert the NAND-NAND logic circuit shown in Figure 6-3 to AND-OR logic and draw the equivalent circuit in the space provided.

$$X = AB + \bar{C}$$

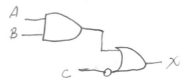

Step 19.   Replace the NAND-NAND logic circuit in Figure 6-3 with the equivalent AND-OR logic circuit in Step 18. Click the On-Off switch to run the simulation. By switching logic switches A, B, and C to the values of A, B, and C in Table 6-3 and monitoring the logic probe light (X), verify the truth table, then stop the simulation.

Questions: Was the AND-OR truth table the same as the NAND-NAND truth table (Table 6-3)? Was the AND-OR logic circuit equivalent to the NAND-NAND logic circuit?

Step 20.   Open circuit file FIG6-4. Write the Boolean equation for the logic circuit in Figure 6-4.

Step 21.   Complete the truth table (Table 6-4) for the logic circuit in Figure 6-4 using the Boolean equation in Step 20. Use the space to the right of the truth table to evaluate each part of the Boolean equation.

Table 6-4 Truth Table

A	B	C	X		
0	0	0			
0	0	1			
0	1	0			
0	1	1			
1	0	0			
1	0	1			
1	1	0			
1	1	1			

Step 22.   Click the On-Off switch to run the simulation. By switching logic switches A, B, and C to the values of A, B, and C in Table 6-4 and monitoring the logic probe light (X), verify the truth table, then stop the simulation.

Question: Was the truth table verified by the logic circuit measurements?

Step 23.   Convert the AND-OR logic circuit in Figure 6-4 to NAND-NAND logic and draw the circuit in the space provided.

Step 24.   Replace the AND-OR logic circuit in Figure 6-4 with the equivalent NAND-NAND logic circuit in Step 23. Click the On-Off switch to run the simulation. By switching logic switches A, B, and C to the values of A, B, and C in Table 6-4 and monitoring the logic probe light (X), verify the truth table, then stop the simulation.

Questions: Was the NAND-NAND truth table the same as the AND-OR truth table (Table 6-4)? Was the NAND-NAND logic circuit equivalent to the AND-OR logic circuit?

EXPERIMENT

7 Simplifying Combinational Logic Circuits

Objectives:

1. Simplifying combinational logic circuits using the Boolean theorems.
2. Simplifying combinational logic circuits using DeMorgan's theorems.

Materials:

One 5 V dc power supply
Three logic switches
One logic probe light
Three two-input AND gates (1-7408 IC)
Two two-input OR gates (1-7432 IC)
Two INVERTERS (1-7404 IC)
One two-input NAND gate (1-7400 IC)
One two-input NOR gate (1-7402 IC)
Three 1 kΩ resistors

Theory:

Review the Theory section of Experiment 6 for writing the Boolean equation for a combinational logic circuit. Review the Theory section of Experiment 4 for the Boolean theorems and DeMorgan's theorems. In this experiment, you will write the Boolean equations for the logic circuits in Figures 7-1, 7-2, 7-3, 7-4, 7-5, and 7-6. Then you will simplify the Boolean equations using the Boolean theorems and DeMorgan's theorems and draw the simplified logic circuits from the simplified Boolean equations. Next, you will compare the truth tables for the simplified logic circuits with the truth tables for the original circuits.

Figure 7-1 AND-OR Combinational Logic Circuit

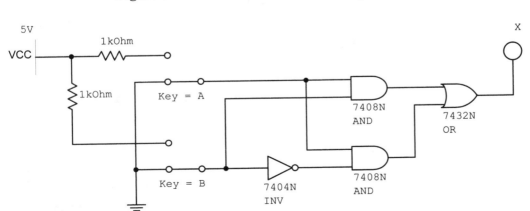

Figure 7-2 AND-OR Combinational Logic Circuit

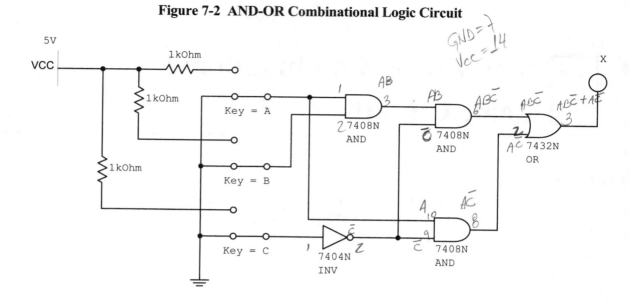

Figure 7-3 OR-AND Combinational Logic Circuit

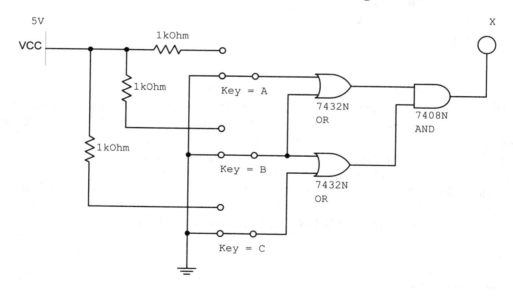

Figure 7-4 Combinational Logic Circuit

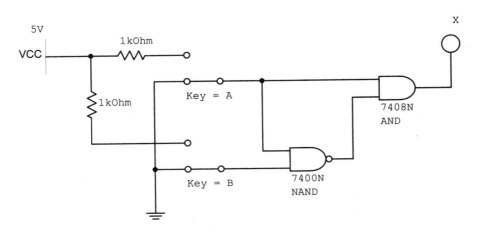

Figure 7-5   Combinational Logic Circuit

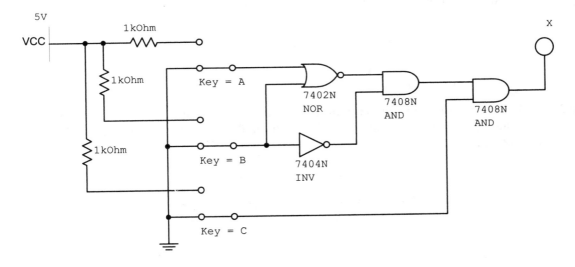

Figure 7-6 Combinational Logic Circuit

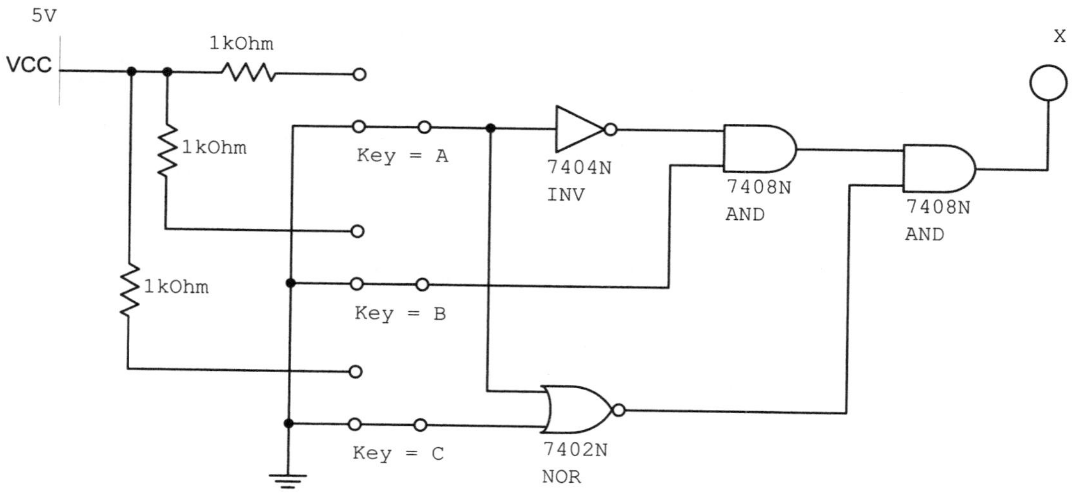

Procedure:

Step 1.   Open circuit file FIG7-1. Write the Boolean equation for the logic circuit in Figure 7-1.

Step 2.   Click the On-Off switch to run the simulation. By switching logic switches A and B to the values of A and B in Table 7-1 and monitoring the logic probe light (X), fill in the truth table, then stop the simulation.

Table 7-1 Truth Table

A	B	X
0	0	
0	1	
1	0	
1	1	

Step 3.   Simplify the Boolean equation in Step 1 using the Boolean theorems.

Step 4.   Draw the simplified logic circuit from the simplified Boolean equation in Step 3.

Step 5.   Replace the logic circuit in Figure 7-1 with the simplified logic circuit in Step 4. Click the On-Off switch to run the simulation. By switching logic switches A and B to the values of A and B in Table 7-2 and monitoring the logic probe light (X), fill in the truth table, then stop the simulation.

Table 7-2 Truth Table

A	B	X
0	0	
0	1	
1	0	
1	1	

Questions: How does the truth table for the simplified logic circuit compare with the truth table for the original circuit (Table 7-1)? Is the simplified logic circuit equivalent to the original logic circuit?

Step 6.   Open circuit file FIG7-2. Write the Boolean equation for the logic circuit in Figure 7-2.
$$X = AB\bar{C} + A\bar{C}$$

Step 7.   Click the On-Off switch to run the simulation. By switching logic switches A, B, and C to the values of A, B, and C in Table 7-3 and monitoring the logic probe light (X), fill in the truth table, then stop the simulation.

Table 7-3 Truth Table

A	B	C	X
0	0	0	0
0	0	1	0
0	1	0	0
0	1	1	0
1	0	0	1
1	0	1	0
1	1	0	1
1	1	1	0

Step 8.   Simplify the Boolean equation in Step 6 using the Boolean theorems.

$$X = AB\bar{C} + A\bar{C}$$
$$= A\bar{C}(B+1)$$
$$= A\bar{C}$$

Step 9.   Draw the simplified logic circuit from the simplified Boolean equation in Step 8.

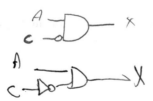

Step 10.   Replace the logic circuit in Figure 7-2 with the simplified logic circuit in Step 9. Click the On-Off switch to run the simulation. By switching logic switches A, B, and C to the values of A, B, and C in Table 7-4 and monitoring the logic probe light (X), fill in the truth table, then stop the simulation.

Table 7-4 Truth Table

A	B	C	X
0	0	0	0
0	0	1	/
0	1	0	/
0	1	1	0
1	0	0	1
1	0	1	/
1	1	0	1
1	1	1	0

Questions: How does the truth table for the simplified logic circuit compare with the truth table for the original circuit (Table 7-3)? Is the simplified logic circuit equivalent to the original logic circuit?

Step 11.   Open circuit file FIG7-3. Write the Boolean equation for the logic circuit in Figure 7-3.

Step 12.   Click the On-Off switch to run the simulation. By switching logic switches A, B, and C to the values of A, B, and C in Table 7-5 and monitoring the logic probe light (X), fill in the truth table, then stop the simulation.

Table 7-5 Truth Table

A	B	C	X
0	0	0	
0	0	1	
0	1	0	
0	1	1	
1	0	0	
1	0	1	
1	1	0	
1	1	1	

Step 13.   Simplify the Boolean equation in Step 11 using the Boolean theorems. Write your answer in sum-of-products (AND-OR) form.

Step 14.   Draw the simplified logic circuit from the simplified Boolean equation in Step 13.

Step 15.   Replace the logic circuit in Figure 7-3 with the simplified logic circuit in Step 14. Click the On-Off switch to run the simulation. By switching logic switches A, B, and C to the values of A, B, and C in Table 7-6 and monitoring the logic probe light (X), fill in the truth table, then stop the simulation.

Table 7-6 Truth Table

A	B	C	X
0	0	0	
0	0	1	
0	1	0	
0	1	1	
1	0	0	
1	0	1	
1	1	0	
1	1	1	

Questions: How does the truth table for the simplified logic circuit compare with the truth table for the original circuit (Table 7-5)? Is the simplified logic circuit equivalent to the original logic circuit?

Step 16.   Open circuit file FIG7-4. Write the Boolean equation for the logic circuit in Figure 7-4.

Step 17.   Click the On-Off switch to run the simulation. By switching logic switches A and B to the values of A and B in Table 7-7 and monitoring the logic probe light (X), fill in the truth table, then stop the simulation.

Table 7-7 Truth Table

A	B	X
0	0	
0	1	
1	0	
1	1	

Step 18.   Simplify the Boolean equation in Step 16 using the Boolean theorems and DeMorgan's theorem.

Step 19.   Draw the simplified logic circuit from the Boolean equation in Step 18.

Step 20.   Replace the logic circuit in Figure 7-4 with the simplified logic circuit in Step 19. Click the On-Off switch to run the simulation. By switching logic switches A and B to the values of A and B in Table 7-8 and monitoring the logic probe light (X), fill in the truth table, then stop the simulation.

Table 7-8 Truth Table

A	B	X
0	0	
0	1	
1	0	
1	1	

Questions: How does the truth table for the simplified logic circuit compare with the truth table for the original circuit (Table 7-7)? Is the simplified logic circuit equivalent to the original logic circuit?

Step 21.   Open circuit file FIG7-5. Write the Boolean equation for the logic circuit in Figure 7-5.

Step 22.   Click the On-Off switch to run the simulation. By switching logic switches A, B, and C to the values of A, B, and C in Table 7-9 and monitoring the logic probe light (X), fill in the truth table, then stop the simulation.

Table 7-9 Truth Table

A	B	C	X
0	0	0	
0	0	1	
0	1	0	
0	1	1	
1	0	0	
1	0	1	
1	1	0	
1	1	1	

Step 23.   Simplify the Boolean equation in Step 21 using the Boolean theorems and DeMorgan's theorems.

Step 24.   Draw the simplified logic circuit from the simplified Boolean equation in Step 23.

Step 25.   Replace the logic circuit in Figure 7-5 with the simplified logic circuit in Step 24. Click the On-Off switch to run the simulation. By switching logic switches A, B, and C to the values of A, B, and C in Table 7-10 and monitoring the logic probe light (X), fill in the truth table, then stop the simulation.

Table 7-10 Truth Table

A	B	C	X
0	0	0	
0	0	1	
0	1	0	
0	1	1	
1	0	0	
1	0	1	
1	1	0	
1	1	1	

Questions: How does the truth table for the simplified logic circuit compare with the truth table for the original circuit (Table 7-9)? Is the simplified logic circuit equivalent to the original logic circuit?

Step 26.   Open circuit file FIG7-6. Write the Boolean equation for the logic circuit in Figure 7-6.

Step 27.   Click the On-Off switch to run the simulation. By switching logic switches A, B, and C to the values of A, B, and C in Table 7-11 and monitoring the logic probe light (X), fill in the truth table, then stop the simulation.

Table 7-11 Truth Table

A	B	C	X
0	0	0	
0	0	1	
0	1	0	
0	1	1	
1	0	0	
1	0	1	
1	1	0	
1	1	1	

Step 28.   Simplify the Boolean equation in Step 26 using the Boolean theorems and DeMorgan's theorems.

Step 29.   Draw the simplified logic circuit from the simplified Boolean equation in Step 28.

Step 30.   Replace the logic circuit in Figure 7-6 with the simplified logic circuit in Step 29. Click the On-Off switch to run the simulation. By switching logic switches A, B, and C to the values of A, B, and C in Table 7-12 and monitoring the logic probe light (X), fill in the truth table, then stop the simulation.

Table 7-12 Truth Table

A	B	C	X
0	0	0	
0	0	1	
0	1	0	
0	1	1	
1	0	0	
1	0	1	
1	1	0	
1	1	1	

Questions: How does the truth table for the simplified logic circuit compare with the truth table for the original circuit (Table 7-11)? Is the simplified logic circuit equivalent to the original logic circuit?

EXPERIMENT

Logic Simplification Using Karnaugh Maps

Objectives:

1.   Simplify AND-OR combinational logic circuits using K-maps.

Materials:

One 5 V dc power supply
Four logic switches
One logic probe light
Seven two-input AND gates (2-7408 IC)
Two two-input OR gates (1-7432 IC)
Three INVERTERS (1-7404 IC)
Four 1 kΩ resistors

Theory:

Review the Theory section of Experiment 6 on writing Boolean equations for combinational logic circuits. In this experiment, you will write the Boolean equations for the AND-OR logic circuits in Figures 8-1, 8-2, 8-3, 8-4, 8-5, and 8-6. Then you will simplify the Boolean equations using Karnaugh maps and draw the simplified AND-OR logic circuits from the simplified Boolean equations. Karnaugh maps provide a cookbook method of simplifying AND-OR logic circuits and provide more certainty in obtaining the simplest logic circuit.

Figure 8-1 AND-OR Combinational Logic Circuit

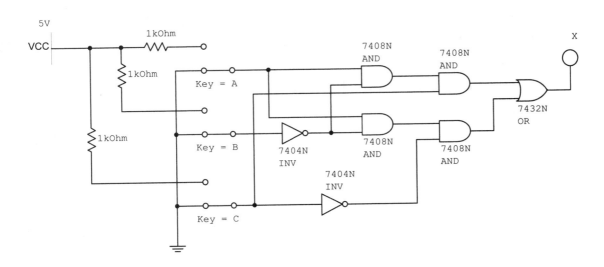

Figure 8-2 AND-OR Combinational Logic Circuit

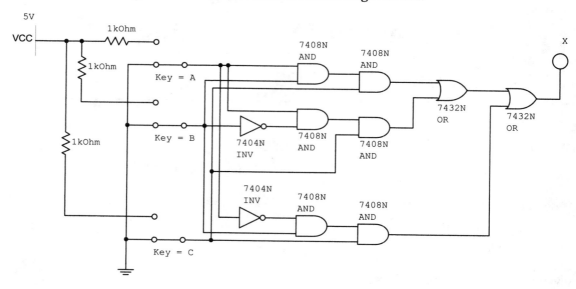

Figure 8-3 AND-OR Combinational Logic Circuit

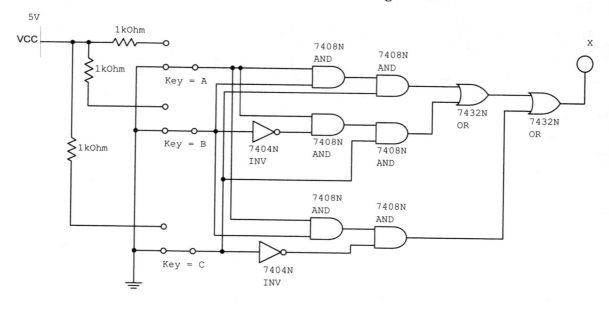

Figure 8-4 AND-OR Combinational Logic Circuit

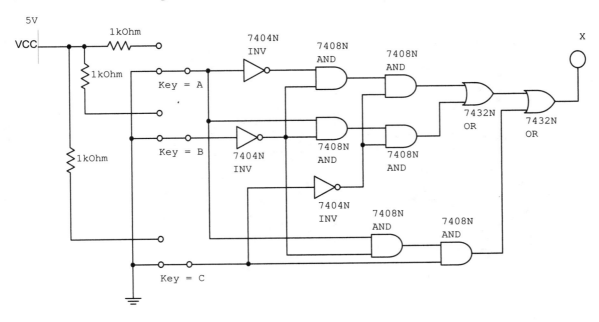

Figure 8-5 AND-OR Combinational Logic Circuit

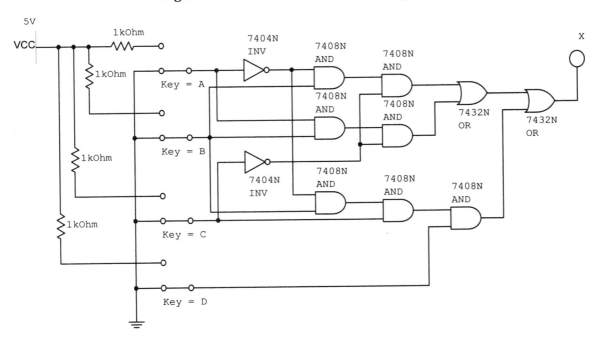

Figure 8-6 AND-OR Combinational Logic Circuit

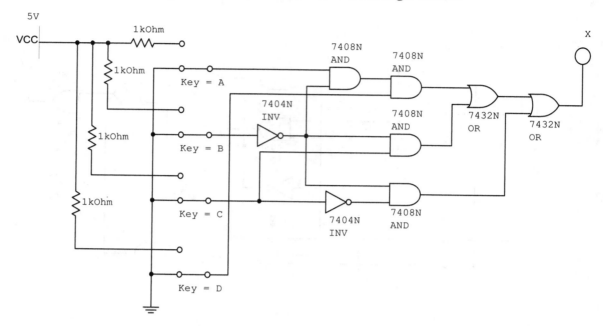

The 2-input K-map is shown in Figure 8-7, the 3-input K-map is shown in Figure 8-8 and the 4-input K-map is shown in Figure 8-9.

Figure 8-7 Two Input K-Map

	$\overline{B}$	B
$\overline{A}$		
A		

Figure 8-8 Three-Input K-map

	$\overline{C}$	C
$\overline{A}\,\overline{B}$		
$\overline{A}B$		
AB		
$A\overline{B}$		

Figure 8-9 Four-Input K-Map

$$\overline{C}\overline{D} \quad \overline{C}D \quad CD \quad C\overline{D}$$

	$\overline{C}\overline{D}$	$\overline{C}D$	CD	$C\overline{D}$
$\overline{A}\,\overline{B}$				
$\overline{A}B$				
AB				
$A\overline{B}$				

When using a **Karnaugh map** (K-map) to simplify a logic circuit you must first write the circuit logic equation in **sum-of-products (AND-OR) form**. Then write a "1" in each K-map block that corresponds to each product term (AND operation) in your logic equation. In some cases the product term will take more than one block in the K-map. This will happen when the number of terms in the product term is less than the number of input variables.

After the K-map is filled with 1's in the appropriate blocks, encircle the adjacent 1's in groups of eight, four, and two, encircling the largest groups first. Consider the sides of the K-map to be adjacent to each other and the top and bottom of the K-map to be adjacent to each other. Overlapping of groups is allowed. Next, encircle any isolated 1's. Make sure that all of the 1's are encircled or within a grouping. **Remove any circles that can be removed without exposing any 1's.**

Each product term in the simplified sum-of-products equation consists of those variables that remain unchanged within each encircled group. Add any product terms representing isolated 1's. These product terms cannot be simplified.

Procedure:

Step 1.   Open circuit file FIG8-1. Write the Boolean equation for the logic circuit in Figure 8-1.

Step 2.   Click the On-Off switch to run the simulation. By switching logic switches A, B, and C to the values of A, B, and C in Table 8-1 and monitoring the logic probe light (X), fill in the truth table, then stop the simulation.

Table 8-1 Truth Table

A	B	C	X
0	0	0	
0	0	1	
0	1	0	
0	1	1	
1	0	0	
1	0	1	
1	1	0	
1	1	1	

Step 3.   Draw the 3-input K-map and use it to simplify the Boolean equation in Step 1. Write the simplified Boolean equation.

Step 4.   Draw the simplified logic circuit from the simplified Boolean equation in Step 3.

Step 5.   Replace the logic circuit in Figure 8-1 with the simplified logic circuit in Step 4. Click the On-Off switch to run the simulation. By switching logic switches A, B, and C to the values of A, B, and C in Table 8-2 and monitoring the logic probe light (X), fill in the truth table, then stop the simulation.

Table 8-2 Truth Table

A	B	C	X
0	0	0	
0	0	1	
0	1	0	
0	1	1	
1	0	0	
1	0	1	
1	1	0	
1	1	1	

Questions: How does the truth table for the simplified logic circuit compare with the truth table for the original circuit (Table 8-1)? Is the simplified logic circuit equivalent to the original logic circuit?

Step 6.   Open circuit file FIG8-2. Write the Boolean equation for the logic circuit in Figure 8-2.

Step 7.   Click the On-Off switch to run the simulation. By switching logic switches A, B, and C to the values of A, B, and C in Table 8-3 and monitoring the logic probe light (X), fill in the truth table, then stop the simulation.

Table 8-3 Truth Table

A	B	C	X
0	0	0	
0	0	1	
0	1	0	
0	1	1	
1	0	0	
1	0	1	
1	1	0	
1	1	1	

Step 8.   Draw the 3-input K-map and use it to simplify the Boolean equation in Step 6. Write the simplified Boolean equation.

Step 9.   Draw the simplified logic circuit from the simplified Boolean equation in Step 8.

Step 10.   Replace the logic circuit in Figure 8-2 with the simplified logic circuit in Step 9. Click the On-Off switch to run the simulation. By switching logic switches A, B, and C to the values of A, B, and C in Table 8-4 and monitoring the logic probe light (X), fill in the truth table, then stop the simulation.

Table 8-4 Truth Table

A	B	C	X
0	0	0	
0	0	1	
0	1	0	
0	1	1	
1	0	0	
1	0	1	
1	1	0	
1	1	1	

Questions: How does the truth table for the simplified logic circuit compare with the truth table for the original circuit (Table 8-3)? Is the simplified logic circuit equivalent to the original logic circuit?

Step 11.   Open circuit file FIG8-3. Write the Boolean equation for the logic circuit in Figure 8-3.

Step 12.   Click the On-Off switch to run the simulation. By switching logic switches A, B, and C to the values of A, B, and C in Table 8-5 and monitoring the logic probe light (X), fill in the truth table, then stop the simulation.

Table 8-5 Truth Table

A	B	C	X
0	0	0	
0	0	1	
0	1	0	
0	1	1	
1	0	0	
1	0	1	
1	1	0	
1	1	1	

Step 13.   Draw the 3-input K-map and use it to simplify the Boolean equation in Step 11. Write the simplified Boolean equation.

Step 14.   Draw the simplified logic circuit from the simplified Boolean equation in Step 13.

Step 15.   Replace the logic circuit in Figure 8-3 with the simplified logic circuit in Step 14. Click the On-Off switch to run the simulation. By switching logic switches A, B, and C to the values of A, B, and C in Table 8-6 and monitoring the logic probe light (X), fill in the truth table, then stop the simulation.

Table 8-6 Truth Table

A	B	C	X
0	0	0	
0	0	1	
0	1	0	
0	1	1	
1	0	0	
1	0	1	
1	1	0	
1	1	1	

Questions: How does the truth table for the simplified logic circuit compare with the truth table for the original circuit (Table 8-5)? Is the simplified logic circuit equivalent to the original logic circuit?

Step 16.   Open circuit file FIG8-4. Write the Boolean equation for the logic circuit in Figure 8-4.

Step 17.   Click the On-Off switch to run the simulation. By switching logic switches A, B, and C to the values of A, B, and C in Table 8-7 and monitoring the logic probe light (X), fill in the truth table, then stop the simulation.

Table 8-7 Truth Table

A	B	C	X
0	0	0	
0	0	1	
0	1	0	
0	1	1	
1	0	0	
1	0	1	
1	1	0	
1	1	1	

Step 18.   Draw the 3-input K-map and use it to simplify the Boolean equation in Step 16. Write the simplified Boolean equation.

Step 19.   Draw the simplified logic circuit from the simplified Boolean equation in Step 18.

Step 20.   Replace the logic circuit in Figure 8-4 with the simplified logic circuit in Step 19. Click the On-Off switch to run the simulation. By switching logic switches A, B, and C to the values of A, B, and C in Table 8-8 and monitoring the logic probe light (X), fill in the truth table, then stop the simulation.

Table 8-8 Truth Table

A	B	C	X
0	0	0	
0	0	1	
0	1	0	
0	1	1	
1	0	0	
1	0	1	
1	1	0	
1	1	1	

Questions: How does the truth table for the simplified logic circuit compare with the truth table for the original circuit (Table 8-7)? Is the simplified logic circuit equivalent to the original logic circuit?

Step 21.   Open circuit file FIG8-5. Write the Boolean equation for the logic circuit in Figure 8-5.

Step 22.   Click the On-Off switch to run the simulation. By switching logic switches A, B, C, and D to the values of A, B, C, and D in Table 8-9 and monitoring the logic probe light (X), fill in the truth table, then stop the simulation.

Table 8-9 Truth Table

A	B	C	D	X
0	0	0	0	
0	0	0	1	
0	0	1	0	
0	0	1	1	
0	1	0	0	
0	1	0	1	
0	1	1	0	
0	1	1	1	
1	0	0	0	
1	0	0	1	
1	0	1	0	
1	0	1	1	
1	1	0	0	
1	1	0	1	
1	1	1	0	
1	1	1	1	

Step 23.   Draw the 4-input K-map and use it to simplify the Boolean equation in Step 21. Write the simplified Boolean equation.

Step 24.   Draw the simplified logic circuit from the simplified Boolean equation in Step 23.

Step 25.   Replace the logic circuit in Figure 8-5 with the simplified logic circuit in Step 24. Click the
            On-Off switch to run the simulation. By switching logic switches A, B, C, and D to the
            values of A, B, C, and D in Table 8-10 and monitoring the logic probe light (X), fill in the
            truth table, then stop the simulation.

Table 8-10 Truth Table

A	B	C	D	X
0	0	0	0	
0	0	0	1	
0	0	1	0	
0	0	1	1	
0	1	0	0	
0	1	0	1	
0	1	1	0	
0	1	1	1	
1	0	0	0	
1	0	0	1	
1	0	1	0	
1	0	1	1	
1	1	0	0	
1	1	0	1	
1	1	1	0	
1	1	1	1	

Questions: How does the truth table for the simplified logic circuit compare with the truth table for the original circuit (Table 8-9)? Is the simplified logic circuit equivalent to the original logic circuit?

Step 26.   Open circuit file FIG8-6. Write the Boolean equation for the logic circuit in Figure 8-6.

Step 27.   Click the On-Off switch to run the simulation. By switching logic switches A, B, C, and D to the values of A, B, C, and D in Table 8-11 and monitoring the logic probe light (X), fill in the truth table, then stop the simulation.

Table 8-11 Truth Table

A	B	C	D	X
0	0	0	0	
0	0	0	1	
0	0	1	0	
0	0	1	1	
0	1	0	0	
0	1	0	1	
0	1	1	0	
0	1	1	1	
1	0	0	0	
1	0	0	1	
1	0	1	0	
1	0	1	1	
1	1	0	0	
1	1	0	1	
1	1	1	0	
1	1	1	1	

Step 28.   Draw the 4-input K-map and use it to simplify the Boolean equation in Step 26. Write the simplified Boolean equation.

Step 29.   Draw the simplified logic circuit from the simplified Boolean equation in Step 28.

Step 30.   Replace the logic circuit in Figure 8-6 with the simplified logic circuit in Step 29. Click the On-Off switch to run the simulation. By switching logic switches A, B, C, and D to the values of A, B, C, and D in Table 8-12 and monitoring the logic probe light (X), fill in the truth table, then stop the simulation.

Table 8-12 Truth Table

A	B	C	D	X
0	0	0	0	
0	0	0	1	
0	0	1	0	
0	0	1	1	
0	1	0	0	
0	1	0	1	
0	1	1	0	
0	1	1	1	
1	0	0	0	
1	0	0	1	
1	0	1	0	
1	0	1	1	
1	1	0	0	
1	1	0	1	
1	1	1	0	
1	1	1	1	

Questions: How does the truth table for the simplified logic circuit compare with the truth table for the original circuit (Table 8-11)? Is the simplified logic circuit equivalent to the original logic circuit?

EXPERIMENT

Designing Combinational Logic Circuits

Objectives:

1. Obtain experience developing truth tables from problem statements.
2. Obtain experience writing sum-of-product (AND-OR) logic expressions from truth tables.
3. Obtain experience designing the simplest NAND-NAND combinational logic circuits from logic expressions.
4. Test combinational logic circuit designs for conformity with the original problem statements or truth tables.

Materials:

One 5 V dc power supply
Four logic switches
One logic probe light
Two-input NAND gates (1-7400 IC)
Three-input NAND gates (1-7410 IC)
1 kΩ resistors

Theory:

In this experiment, you will design several **combinational logic circuits** from problem statements. First, you will develop a **truth table** from the problem statement, write the **AND-OR logic equation** from the truth table, simplify the logic equation using a **K-map**, and draw the **simplified AND-OR logic circuit** from the simplified logic equation. Then you will convert the AND-OR logic circuit to NAND-NAND logic and build and test the NAND-NAND logic circuit. You will build the NAND-NAND logic circuit on the computer using Multisim or in a hardwired laboratory and determine if the truth table for your circuit design matches the original truth table from the problem statement. Use logic switches to control the logic inputs and a logic probe light to monitor the output.

A truth table lists the outputs for all of the possible input combinations for a logic circuit. A sum-of-products (AND-OR) Boolean equation can be derived from a truth table by determining the input combinations required to produce a binary "1" output and representing each of these input combinations as an AND operation. The Boolean equation is determined by ORing all of the AND operations, resulting in an AND-OR logic equation.

Review the Theory section of Experiment 8 for information on using K-maps to simplify logic equations. Review the Theory section of Experiment 6 for information on how to convert AND-OR logic to NAND-NAND logic.

Procedure:

Step 1:   Develop the truth table for a 2-input combinational logic circuit whose output is high only
            when one of the inputs is high.

Step 2.   Write the sum-of-products (AND-OR) Boolean equation for the truth table in Step 1.

Step 3.   Using a K-map, simplify the Boolean equation in Step 2, if possible.

Questions: Were you able to simplify the Boolean equation? Was the original equation already in its
simplest form?

Step 4.   Draw the simplest AND-OR logic circuit from the simplest Boolean equation in Step 3.

Step 5.   Draw the equivalent NAND-NAND logic circuit for the AND-OR logic circuit in Step 4.
            Use NAND gates to implement INVERTERS.

Step 6.   Construct the NAND-NAND logic circuit drawn in Step 5. Use logic switches for the logic
            circuit inputs and use a logic probe light to monitor the output for each input combination
            to determine if your circuit truth table matches the truth table from the problem statement.

Question: How did the truth table for your simplified NAND-NAND logic circuit compare with the
truth table from the problem statement in Step 1? Did your logic circuit design satisfy the conditions of
the stated problem?

Step 7.   Develop the truth table for a 3-input combinational logic circuit whose output is high when
            a majority of the inputs are high.

Step 8.   Write the sum-of-products (AND-OR) Boolean equation for the truth table in Step 7.

Step 9.   Using a K-map, simplify the Boolean equation in Step 8, if possible.

Questions: Were you able to simplify the Boolean equation? Was the original equation already in its simplest form?

Step 10.   Draw the simplest AND-OR logic circuit from the simplest Boolean equation in Step 9.

Step 11.   Draw the equivalent NAND-NAND logic circuit for the AND-OR logic circuit in Step 10. Use NAND gates to implement INVERTERS.

Step 12.   Construct the NAND-NAND logic circuit drawn in Step 11. Use logic switches for the logic circuit inputs and use a logic probe light to monitor the output for each input combination to determine if your circuit truth table matches the truth table from the problem statement.

Questions: How did the truth table for your simplified NAND-NAND logic circuit compare with the truth table from the problem statement in Step 7? Did your logic circuit design satisfy the conditions of the stated problem?

Step 13.   Develop the truth table for a 3-input combinational logic circuit whose output is the inverse of input A when inputs B and C are equal, and whose output is high when inputs B and C are different.

Step 14.   Write the sum-of-products (AND-OR) Boolean equation for the truth table in Step 13.

Step 15.   Using a K-map, simplify the Boolean equation in Step 14, if possible.

Questions: Were you able to simplify the Boolean equation? Was the original equation already in its simplest form?

Step 16.   Draw the simplest AND-OR logic circuit from the simplest Boolean equation in Step 15.

Step 17.   Draw the equivalent NAND-NAND logic circuit for the AND-OR logic circuit in Step 16. Use NAND gates to implement INVERTERS.

Step 18.   Construct the NAND-NAND logic circuit drawn in Step 17. Use logic switches for the logic circuit inputs and use a logic probe light to monitor the output for each input combination to determine if your circuit truth table matches the truth table from the problem statement.

Questions: How did the truth table for your simplified NAND-NAND logic circuit compare with the truth table from the problem statement in Step 13? Did your logic circuit design satisfy the conditions of the stated problem?

Step 19.   Develop the truth table for a 4-input combinational logic circuit that will produce a high
            output when the 4-bit binary input is greater than 9, and will produce a low output when the
            4-bit binary input is 9 or less.

Step 20.   Write the sum-of-products (AND-OR) Boolean equation for the truth table in Step 19.

Step 21.   Using a K-map, simplify the Boolean equation in Step 20, if possible.

Questions: Were you able to simplify the Boolean equation? Was the original equation already in its
simplest form?

Step 22.   Draw the simplest AND-OR logic circuit from the simplest Boolean equation in Step 21.

Step 23.   Draw the equivalent NAND-NAND logic circuit for the AND-OR logic circuit in Step 22.
             Use NAND gates to implement INVERTERS.

Step 24.   Construct the NAND-NAND logic circuit drawn in Step 23. Use logic switches for the logic
             circuit inputs and use a logic probe light to monitor the output for each input combination to
             determine if your circuit truth table matches the truth table from the problem statements.

Questions: How did the truth table for your simplified NAND-NAND logic circuit compare with the
truth table from the problem statement in Step 19? Did your logic circuit design satisfy the conditions
of the stated problem?

Step 25.   Develop the truth table for a 4-input combinational logic circuit that will output a high whenever the 4-bit binary input represents an even number and output a low whenever the 4-bit binary input represents an odd number.

Step 26.   Write the sum-of-products (AND-OR) Boolean equation for the truth table in Step 25.

Step 27.   Using a K-map, simplify the Boolean equation in Step 26, if possible.

Questions: Were you able to simplify the Boolean equation? Was the original equation already in its simplest form?

Step 28.   Draw the simplest AND-OR logic circuit from the simplest Boolean equation in Step 27.

Step 29.   Draw the equivalent NAND-NAND logic circuit for the AND-OR logic circuit in Step 28. Use NAND gates to implement INVERTERS.

Step 30.   Construct the NAND-NAND logic circuit drawn in Step 29. Use logic switches for the logic circuit inputs and use a logic probe light to monitor the output for each input combination to determine if your circuit truth table matches the truth table from the problem statement.

Questions: How did the truth table for your simplified NAND-NAND logic circuit compare with the truth table from the problem statement in Step 25? Did your logic circuit design satisfy the conditions of the stated problem?

EXPERIMENT

10 Troubleshooting Combinational Logic Circuits

Objectives:

1. Determine the defective logic gates for various combinational logic circuits by monitoring the logic levels at circuit test points.

Materials:

This experiment can only be performed on Electronics Workbench Multisim using the circuits disk provided with this manual.

Theory:

In order to perform this experiment effectively, you must first complete Experiments 1–6. Use the theory learned in those experiments to find the defective logic gates in the logic circuits in this experiment.

Determine the defective gate in each experiment by trying all possible logic network binary input combinations and noting the binary inputs and outputs on each logic gate. A logic gate is defective if it has an incorrect output for any combination of binary inputs. Remember that an open input on a 7400 series logic gate will behave as if there is a logical "one" on that open input terminal, even if that open input is being caused by the open output of another logic gate. For example, if one of the inputs of an OR gate that is not defective is connected to a logic gate with an open output, the OR gate will act as if there is a logical "one" on that input. For this reason, make sure that you try all possible logic network binary input combinations before concluding which logic gate is defective.

Procedure:

Don't forget to read the Theory section before attempting to determine the defective gates in the following logic circuits.

1. Open circuit file FIG10-1. Click the On-Off switch to run the simulation. Based on the logic levels at the test points for various logic network binary inputs, determine which logic gate is defective. To switch the logic switches, press the key on the computer keyboard that matches the letter label on the switch.

    Defective gate _U1A - 7408N_

2.   Open circuit file FIG10-2. Click the On-Off switch to run the simulation. Based on the logic levels at the test points for various logic network binary inputs, determine which logic gate is defective. To switch the logic switches, press the key on the computer keyboard that matches the letter label on the switch.

Defective gate   U1A - 7408A

3.   Open circuit file FIG10-3. Click the On-Off switch to run the simulation. Based on the logic levels at the test points for various logic network binary inputs, determine which logic gate is defective. To switch the logic switches, press the key on the computer keyboard that matches the letter label on the switch.

Defective gate   U1B - 7408A

4.   Open circuit file FIG10-4. Click the On-Off switch to run the simulation. Based on the logic levels at the test points for various logic network binary inputs, determine which logic gate is defective. To switch the logic switches, press the key on the computer keyboard that matches the letter label on the switch.

Defective gate   U2A - 7432N

5.   Open circuit file FIG10-5. Click the On-Off switch to run the simulation. Based on the logic levels at the test points for various logic network binary inputs, determine which logic gate is defective. To switch the logic switches, press the key on the computer keyboard that matches the letter label on the switch.

Defective gate   U3B - 7404N

6.   Open circuit file FIG10-6. Click the On-Off switch to run the simulation. Based on the logic levels at the test points for various logic network binary inputs, determine which logic gate is defective. To switch the logic switches, press the key on the computer keyboard that matches the letter label on the switch.

Defective gate   U1B - 7400N

7.   Open circuit file FIG10-7. Click the On-Off switch to run the simulation. Based on the logic levels at the test points for various logic network binary inputs, determine which logic gate is defective. To switch the logic switches, press the key on the computer keyboard that matches the letter label on the switch.

Defective gate   U1A - 7408N

8.   Open circuit file FIG10-8. Click the On-Off switch to run the simulation. Based on the logic levels at the test points for various logic network binary inputs, determine which logic gate is defective. To switch the logic switches, press the key on the computer keyboard that matches the letter label on the switch.

Defective gate   U3A - 7404N

9.   Open circuit file FIG10-9. Click the On-Off switch to run the simulation. Based on the logic levels at the test points for various logic network binary inputs, determine which logic gate is defective. To switch the logic switches, press the key on the computer keyboard that matches the letter label on the switch.

Defective gate _U1B - 7408N_

10.   Open circuit file FIG10-10. Click the On-Off switch to run the simulation. Based on the logic levels at the test points for various logic network binary inputs, determine which logic gate is defective. To switch the logic switches, press the key on the computer keyboard that matches the letter label on the switch.

Defective gate _U4A - 7408N_

11.   Open circuit file FIG10-11. Click the On-Off switch to run the simulation. Based on the logic levels at the test points for various logic network binary inputs, determine which logic gate is defective. To switch the logic switches, press the key on the computer keyboard that matches the letter label on the switch.

Defective gate _U2A - 7432N_

12.   Open circuit file FIG10-12. Click the On-Off switch to run the simulation. Based on the logic levels at the test points for various logic network binary inputs, determine which logic gate is defective. To switch the logic switches, press the key on the computer keyboard that matches the letter label on the switch.

Defective gate _U3B - 7404N_

II

Arithmetic Logic Circuits

The experiments in Part II involve the study of **arithmetic logic circuits**. You will study XOR and XNOR gates, and then determine how they are used to build **half-adders, full-adders, binary adders, BCD adders, parity generators and checkers**, and **magnitude comparators**. In the final experiment in Part II you will solve some **troubleshooting problems** in arithmetic logic circuits.

The circuits for the experiments in Part II can be found on the enclosed disk in the PART2 subdirectory.

Name_____

Date_____

Logic Gates: XOR and XNOR

Objectives:

1.   Complete the truth table for an XOR gate and compare it with the OR gate truth table.
2.   Plot the XOR gate pulse response timing diagram.
3.   Complete the truth table for an XNOR gate and compare it with the XOR gate truth table.
4.   Plot the XNOR gate pulse response timing diagram.
5.   Demonstrate the use of an XOR gate as a controlled inverter.
6.   Design the simplest AND-OR logic network that satisfies the XOR truth table.
7.   Design the simplest AND-OR logic network that satisfies the XNOR truth table.

Materials:

One 5 V dc voltage supply
Two logic switches
Three logic probe lights
One logic analyzer
Two pulse generators
One XOR gate (1-7486 IC)
INVERTERS (1-7404 IC)
Two-input AND gates (1-7408 IC)
Two-input OR gates (1-7432 IC)
Resistors—1 kΩ (2), 10 kΩ

Theory:

A **truth table** lists the outputs for all of the possible input combinations for a logic gate or logic network. See the results in Experiment 3 for the OR gate truth table.

The **XOR gate** output will be high (1) when only one of the inputs is high (1). The XOR gate output will be low (0) when both of the inputs are low (0) or high (1). A circuit for studying the XOR gate is shown in Figure 11-1. The circuit for plotting the **XOR gate pulse response timing diagram** is shown in Figure 11-2.

An XNOR gate is an XOR gate with an INVERTER on the output. The output will be high (1) when both of the inputs are low (0) or high (1). The output will be low (0) when only one of the inputs is high (1). A circuit for studying the XNOR gate is shown in Figure 11-3. The circuit for plotting the **XNOR gate pulse response timing diagram** is shown in Figure 11-4.

If one of the inputs of an XOR gate is connected to a logic switch, as shown in Figure 11-5, it can be used as a **controlled inverter**. The XOR gate output (Y) will be equal to the other XOR gate input (A) when the logic switch is low (0), and the XOR gate output (Y) will be the inverse of the other XOR gate input (A) when the logic switch is high (1). This can be demonstrated from the XOR gate truth table.

Complete Experiment 9 before designing the XOR and XNOR logic circuits in Steps 7–12. If you have already completed this experiment, you may want to review the Theory section.

Figure 11-1 XOR Gate

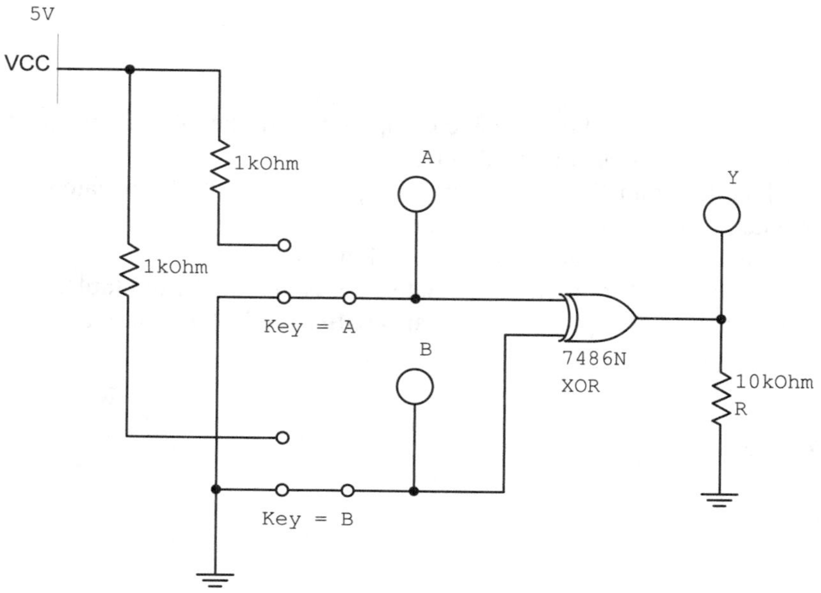

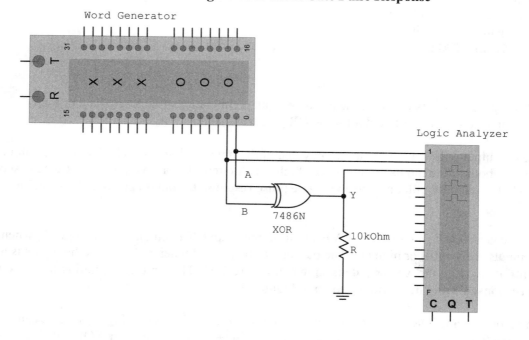

Figure 11-2 XOR Gate Pulse Response

Figure 11-3 XNOR Gate

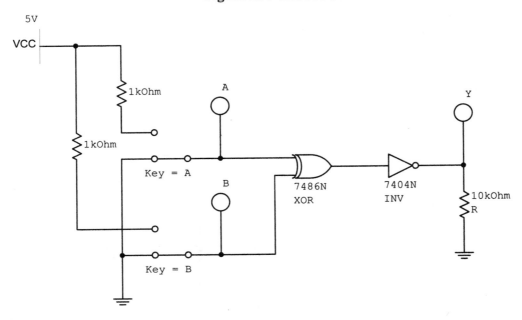

Figure 11-4 XNOR Gate Pulse Response

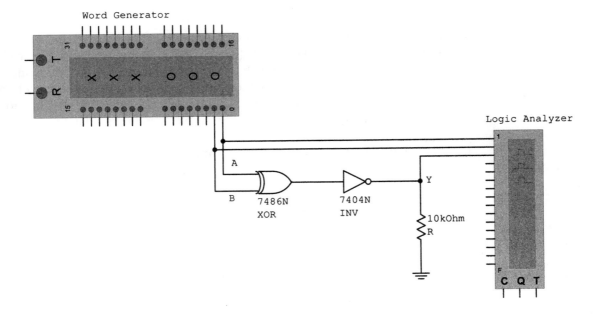

Figure 11-5 XOR Gate Controlled Inverter

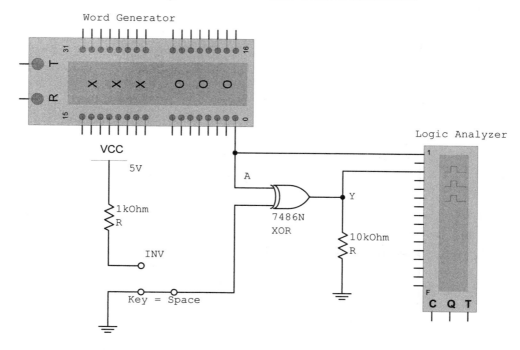

Procedure:

Step 1.   Open circuit file FIG11-1. Notice that Logic Switches A and B are placing binary "zeros" (ground equals 0 V) on the two XOR gate inputs (A and B). Resistor R represents the load on the XOR gate output. Click the On-Off switch to run the simulation. By switching the logic switches to the appropriate positions, complete the XOR gate truth table (Table 11-1). You can switch the logic switches by pressing the letter on the keyboard corresponding to the letter label on the switch. After completing the truth table, stop the simulation.

Table 11-1 XOR Gate Truth Table

A	B	Y
0	0	
0	1	
1	0	
1	1	

Questions: Based on the results in Table 11-1, what conclusion can you draw about the relationship between the XOR gate binary output and the binary inputs?

What is the difference between the XOR gate truth table and the OR gate truth table?

Step 2.   Open circuit file FIG 11-2. Bring down the word generator enlargement and make sure that the following settings are selected: Frequency = 1 kHz, Trigger = Internal, Hex codes (00000000, 00000001, 00000002, 00000003, repeated four times). Move the word generator to the left by clicking and dragging. Bring down the logic analyzer enlargement and make sure that the following settings are selected: Clocks/Div = 16, Clock Setup (Clock Source = Internal, Clock Rate = 10 kHz, Pre-trigger Samples = 100, Post-trigger Samples = 1000, Threshold Voltage = 2.5), Trigger Settings (Trigger Clock Edge = Positive, Trigger Qualifier = x, Pattern A= xxxxxxxxxxxxxxxx, Trigger Combinations = A). Click BURST on the word generator to run the simulation. Notice that the word generator has applied a pulse pattern to each XOR gate input, shown in red and green on the logic analyzer screen. The blue curve plot is the XOR gate output. Draw the XOR gate input and output curve plots in the space provided and label them, then stop the simulation.

NOTE: In a hardwired laboratory, use pulse generator outputs to apply square waves to the XOR gate inputs.

Questions: Does the XOR gate output go high only when one of the inputs is high? Is this expected for an XOR gate?

Does the XOR gate output go low only when both inputs are high or low? Is this expected for an XOR gate?

Step 3.   Open circuit file FIG11-3. The XOR gate with an INVERTER on the output is an XNOR
                 gate. Notice that Logic Switches A and B are placing binary zeros (ground equals 0 V) on
                 the two XNOR gate inputs (A and B). Resistor R represents the load on the XNOR gate
                 output. Click the On-Off switch to run the simulation. By switching the logic switches to
                 the appropriate positions, complete the XNOR gate truth table (Table 11-2). When the truth
                 table is complete, stop the simulation.

Table 11-2 XNOR Gate Truth Table

A	B	Y
0	0	
0	1	
1	0	
1	1	

Questions: Based on the results in Table 11-2, what conclusion can you draw about the relationship
between the XNOR gate binary output and the binary inputs?

What is the difference between the XNOR gate truth table and the XOR gate truth table? **Explain.**

Step 4.   Open circuit file FIG 11-4. The word generator and logic analyzer settings should be the
                 same as in Step 2. Click BURST on the word generator to run the simulation. Notice that
                 the word generator has applied a pulse pattern to each XNOR gate input, shown in red and
                 green on the logic analyzer screen. The blue curve plot is the XNOR gate output. Draw the
                 XNOR gate input and output curve plots in the space provided and label them, then stop the
                 simulation.

NOTE: In a hardwired laboratory, use pulse generator outputs to apply square waves to the XNOR gate inputs.

Questions: Does the XNOR gate output go high only when both of the inputs are high or low? Is this expected for an XNOR gate?

Step 5.   Open circuit file FIG11-5. The INV switch should be down (0). The INV switch can be changed by pressing the space bar on the keyboard. Bring down the word generator enlargement and make sure that the following settings are selected: Frequency = 1 kHz, Trigger = Internal, Hex codes (00000000, 00000000, 00000001, 00000001, repeated four times). Move the word generator to the left by clicking and dragging. Bring down the logic analyzer enlargement and make sure that the following settings are selected: Clocks/Div = 16, Clock Setup (Clock Source = Internal, Clock Rate = 10 kHz, Pre-trigger Samples = 100, Post-trigger Samples = 1000, Threshold Voltage = 2.5), Trigger Settings (Trigger Clock Edge = Positive, Trigger Qualifier = x, Pattern A= xxxxxxxxxxxxxxxx, Trigger Combinations = A). Click BURST on the word generator to run the simulation. Notice that the word generator has applied a pulse pattern to the XOR gate input (A), shown in red on the logic analyzer screen. The blue curve plot is the XOR gate output (Y). Draw the input (A) and output (Y) curve plots in the space provided and label them, then stop the simulation.

NOTE: In a hardwired laboratory, use a pulse generator in place of the word generator and an oscilloscope in place of the logic analyzer.

Question: What is the relationship between the input curve plot (A) and the output curve plot (Y)?

Step 6.   Click the arrow in the circuit space, then press the space bar to switch the INV switch to the
             up (1) position. Reset the cursor on the word generator by right clicking the first column
             number and selecting "Set Cursor", if using Multisim 7. Click BURST on the word
             generator to run the simulation again. Draw the input (A) and output (Y) curve plots in the
             space provided and label them, then stop the simulation.

Questions: What is the relationship between the input curve plot (A) and the output curve plot (Y)?

What conclusion can you draw about the relationship between the output (Y) and the input (A) based
on the position of the INV switch?

Step 7.   Based on the XOR gate truth table (Table 11-1) write the simplest AND-OR logic equation
             that will satisfy the truth table.

Step 8.   Based on the logic equation in Step 7, draw the simplest AND-OR logic circuit that
             satisfies the equation.

Step 9.   Construct the AND-OR logic circuit drawn in Step 8. Use logic switches for the logic circuit inputs and use a logic probe light to monitor the output. For each input combination, determine if your circuit matches the XOR truth table in Table 11-1.

Questions: Did the AND-OR logic circuit match the XOR truth table in Table 11-1?

Is your AND-OR logic circuit design equivalent to the XOR gate?

Step 10.   Based on the XNOR gate truth table (Table 11-2) write the simplest AND-OR logic equation that will satisfy the truth table.

Step 11.   Based on the logic equation in Step 10, draw the simplest AND-OR logic circuit that satisfies the equation.

Step 12.   Construct the AND-OR logic circuit drawn in Step 11. Use logic switches for the logic circuit inputs and use a logic probe light to monitor the output. For each input combination, determine if your circuit matches the XNOR truth table in Table 11-2.

Questions: Did the AND-OR logic circuit match the XNOR truth table in Table 11-2?

Is your AND-OR logic circuit design equivalent to the XNOR gate?

EXPERIMENT

12 Arithmetic Circuits

Objectives:

1. Design a binary half-adder using an XOR gate and an AND gate.
2. Design a binary full-adder using two half-adders.
3. Design a binary full-adder using AND-OR logic.

Materials:

One 5 V dc power supply
Three logic switches
Two logic probe lights
XOR gates (1-7486 IC)
Two-input AND gates (1-7408 IC)
Three-input AND gates (2-7411 ICs)
Two-input OR gates (1-7432 IC)
INVERTERS (1-7404 IC)
1 kΩ resistors

Theory:

Before designing the half-adder in Steps 1–5, make sure you complete Experiment 11 on XOR gates. If this experiment has been completed, review the Theory section. You should also review Experiment 2 on AND gates.

A **half-adder** adds two binary input bits (A and B) and produces a **sum output (S)** and a **carry output (C)**. The carry output (C) will go to a binary one (1) when the two input bits (A and B) are both one (1), and the sum output (S) will go to zero. The **truth table for the half-adder** is shown in Table 12-1.

A **full-adder** adds two binary input bits (A and B) and a carry input bit (Cin) and produces a sum output (S) and a carry output (Co). The carry output (Co) will go to a binary one (1) when two or more of the input bits are binary one (1). You can build a binary full-adder from two binary half-adders by connecting the sum output of the first half-adder to one of the inputs of the second half-adder, and connecting the carry input (Cin) to the other input of the second half-adder. This will produce the sum of the three input bits. If either half-adder produces a carry output, then the full-adder should produce a carry output (Co). Therefore, you will need to connect the carry output of each half-adder to a two-input OR gate. The output of the two-input OR gate is the full-adder carry output (Co). A **truth table for a binary full-adder** is shown in Table 12-2.

Before designing the full-adder using AND-OR logic in Steps 11-18, review Experiment 9 on designing combinational logic circuits and Experiment 8 on Karnaugh maps.

Table 12-1 Half-Adder Truth Table

A	B	C	S
0	0	0	0
0	1	0	1
1	0	0	1
1	1	1	0

Table 12-2 Full-Adder Truth Table

A	B	Cin	Co	S
0	0	0	0	0
0	0	1	0	1
0	1	0	0	1
0	1	1	1	0
1	0	0	0	1
1	0	1	1	0
1	1	0	1	0
1	1	1	1	1

Procedure:

Step 1.   The truth table for a binary half-adder is shown in Table 12-1. From the truth table, design a logic circuit to produce the half-adder sum (S) output using an XOR gate, and draw the logic circuit in the space provided.

Step 2.   From the half-adder truth table (Table 12-1), design a logic circuit to produce the carry (C) output using an AND gate and draw the logic circuit in the space provided.

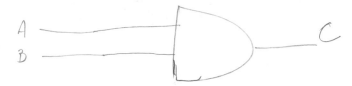

Step 12.   Based on the full-adder truth table (Table 12-2), develop the simplest AND-OR logic equation for the carry output (Co). Use a K-map to simplify the equation, if possible.

Step 13.   Draw the AND-OR logic circuit to satisfy the logic equation developed for the sum output (S) in Step 11.

Step 14.   Draw the simplest AND-OR logic circuit to satisfy the logic equation developed for the
            carry output (Co) in Step 12.

Step 15.   Wire the AND-OR logic circuit for the sum (S) output drawn in Step 13. Use three logic
            switches for inputs A, B, and Cin. Use a logic probe light to monitor the sum output (S).
            You can wire the circuit on the computer using Electronics Workbench Multisim or in a
            hardwired laboratory. If you wire the circuit in a hardwired lab, don't forget to connect +5
            V and ground to each IC chip.

Step 16.   Wire the AND-OR logic circuit for the carry output (Co) drawn in Step 14. Use the same
            three logic switches used in Step 15 for inputs A, B, and Cin. Use a logic probe light to
            monitor the carry output (Co).

Step 17.   Click the On-Off switch to run the simulation (or turn on the power). Compare your sum
            (S) and carry (Co) outputs with the full-adder truth table (Table 12-2) for all combinations
            of the inputs (A, B, and Cin).

Question: Did your full-adder truth table match the truth table in Table 12-2? If not, why?

Step 18.   After your full-adder is working properly, save it on your own personal disk. (You can't
            save any circuits on the disk provided with this manual because it is write protected.)

NOTE: If the circuit was hardwired, save the circuit.

EXPERIMENT

13 Parallel Binary Adder

Objectives:

1. Investigate the operation of a four-bit binary adder.
2. Obtain experience adding binary numbers.
3. Investigate the operation of a four-bit 2's complement adder/subtractor.
4. Obtain experience adding and subtracting binary numbers in the 2's complement number system.

Materials:

One 5 V dc power supply
Nine logic switches
Thirteen logic probe lights
One 4-bit binary parallel adder (1-7483 IC)
Four XOR gates (1-7486 IC)

Theory:

A **4-bit binary adder** consists of four **full-adders** wired as shown in Figure 13-1a. Notice that the **carry output (CARRY)** of each full adder is wired to the **carry input (CIN)** of the next full adder, from right-to-left. Therefore, each full adder will add one bit of the A number to one bit of the B number plus any carry output from the previous full adder, as shown in the following example.

```
     C2 C1 C0  0
     A3 A2 A1 A0
     B3 B2 B1 B0
Cout S3 S2 S1 S0
```

Notice that **binary zero (0 V)** is being applied to the CIN terminal of the first full adder (binary adder carry input), adding 0 to the answer (S). If a binary one (5 V) were applied to the binary adder carry input, 1 would be added to the answer (S).

If two 4-bit binary adders were connected together to produce an **8-bit adder**, the carry output of the first 4-bit binary adder on the right would be connected to the carry input of the second 4-bit binary adder on the left.

If this experiment is performed in a **hardwired laboratory**, a **7483 4-bit binary adder** should be wired in place of the four full adders, as shown in Figure 13-1b. The logic circuit inside the 7483 is similar to the logic shown in Figure 13-1a. Pin 1 on the 7483 is located in the lower left hand corner of the chip, as designated by the dot in Figure 13-1b. See Appendix A for the pin diagram of the 7483 IC chip.

In the **2's complement number system**, the **first bit** of each binary number represents the **sign bit** and determines the sign of the binary number. A **zero (0) sign bit** represents a **positive number** and a **one (1) sign bit** represents a **negative number**. The **2's complement** of a binary number is calculated by inverting each binary bit and then adding a binary one (1). Taking the 2's complement of a binary number reverses the sign of the number, making a positive number negative and a negative number positive. A **negative binary number** is represented by the 2's complement of its positive binary value. This will automatically place a binary one (1) in the sign bit. Therefore, adding a binary number to the 2's complement of another binary number is the equivalent of adding a positive number to a negative number, or the equivalent of **subtracting** one binary number from another binary number. When the answer is negative (binary one in the sign bit), it will be in 2's complement form. The magnitude of the answer will not be apparent until the 2's complement of the negative answer is obtained.

The **4-bit binary adder/subtractor** in Figure 13-2a operates in the **2's complement binary number system** and consists of four full adders and four XOR gates. The four XOR gates are wired as four **controlled inverters**. (See Theory section of Experiment 11.) They are controlled by the SUB switch in Figure 13-2a. When the SUB switch is down (0), the XOR gates let each binary bit of the B number straight through to the full adders. This will cause the 4-bit binary adder to add the A number to the B number. This is the **ADD mode** of the adder/subtractor. When the SUB switch is up (1), it places a binary one (1) on the carry input of the first full adder (4-bit adder carry input), and a binary one (1) on one input of each XOR gate causing the XOR gates to invert each bit of the B number. Inverting each bit of the B number and adding a binary one (1) is the same as taking the 2's complement of the B number. This will cause the 4-bit adder to add the A number to the 2's complement of the B number, which is the equivalent of subtracting the B number from the A number. This is the **SUBTRACT mode** of the adder/subtractor.

If this adder/subtractor experiment is performed in a **hardwired laboratory**, a **7483 4-bit binary adder** should be wired in place of the four full adders, as shown in Figure 13-2b.

Because the 4-bit binary numbers placed on the input of the adder/subtractor will be in the 2's complement number system, the **range of numbers or answers** cannot exceed **decimal +7(0111)** or **decimal −8 (1000)**. If this range of numbers is exceeded, an **overflow** of the number into the sign bit will occur. A larger number of bits is needed to represent a larger range of numbers.

In the **subtraction mode**, the adder/subtractor carry output becomes a **borrow output** and will go to a binary zero (0) when a larger number is subtracted from a smaller number. This represents a borrow from the next column. The carry (borrow) output will be a binary one (1) when a smaller number is subtracted from a larger number and there is no borrow.

Figure 13-1a Four-Bit Binary Adder

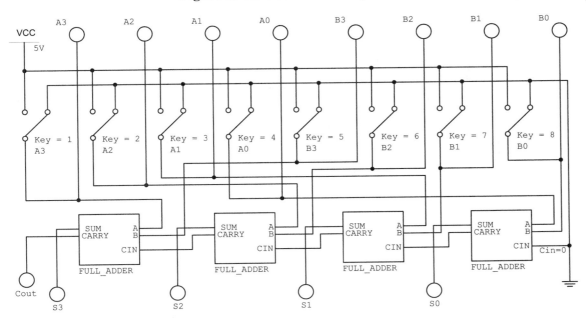

Figure 13-1b 7483 Four-Bit Binary Adder

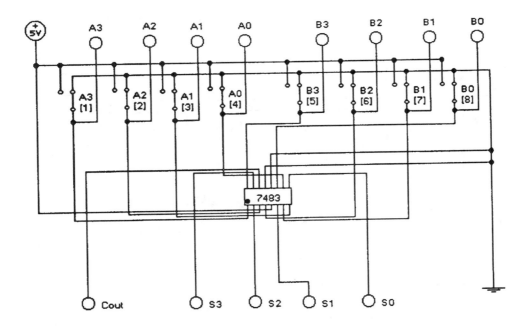

Figure 13-2a Four-Bit Binary Adder/Subtractor

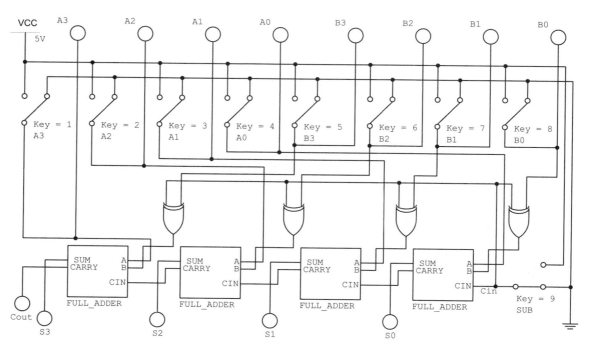

Figure 13-2b 7483 Four-Bit Binary Adder/Subtractor

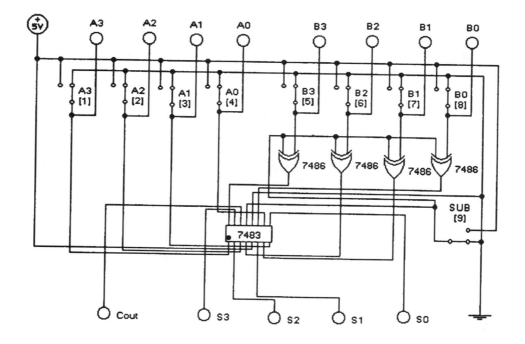

Procedure:

Step 1. Open circuit file FIG13-1. Your are looking at a 4-bit binary adder constructed from four full adders. See the Theory section for more details. You will use the 4-bit binary adder to confirm the results when two 4-bit binary numbers are added. Press number keys 1-8 to control the logic switches. If you are performing this experiment in a **hardwired laboratory**, wire the **7483 4-bit binary adder** as shown in Figure 13-1b.

Step 2. Add binary 0111 to binary 0101 in the space provided and confirm your results using the 4-bit binary adder in Figure 13-1. Also show the decimal equivalent of the numbers and your answer.

Questions: Did your calculated answer match the answer on the adder?

Was the decimal equivalent of your answer correct?

Step 3. Add binary 1111 to binary 1001 in the space provided and confirm your results using the 4-bit binary adder in Figure 13-1. Also show the decimal equivalent of the numbers and your answer.

Questions: Did your calculated answer match the answer on the adder?

Was the decimal equivalent of your answer correct?

Step 4. Convert decimal 11 and decimal 10 to 4-bit binary numbers and add them in the space provided. Confirm your results using the 4-bit binary adder in Figure 13-1.

Questions: Did your calculated answer match the answer of the adder?

Was the decimal equivalent of your answer correct?

Step 5.   Convert decimal 13 and decimal 7 to 4-bit binary numbers and add them in the space
            provided. Confirm your results using the 4-bit binary adder in Figure 13-1.

Questions: Did your calculated answer match the answer on the adder?

Was the decimal equivalent of your answer correct?

Step 6.   Open circuit file FIG13-2. Your are looking at a 4-bit binary adder/subtractor constructed
            from four full adders and four XOR gates. Notice that the four XOR gates are wired as four
            controlled inverters, controlled by the SUB switch. When the SUB switch is down (0), the
            adder/subtractor will add the A number to the B number. When the sub switch is up (1), the
            adder/subtractor will subtract the B number from the A number. Press the 9 key to control
            the SUB switch. See the Theory section for more details. You will use this 4-bit
            adder/subtractor to confirm the results when two 4-bit binary numbers are added or
            subtracted *in the 2's complement number system*. Remember your number and answers
            cannot be outside the range of +7 (0111) to −8 (1000) or the number will overflow into the
            sign bit. If this adder/subtractor experiment is performed in a **hardwired laboratory**, a
            7483 4-bit binary adder should be wired in place of the four full adders, as shown in
            Figure 13-2b.

Step 7.   Add 0110 to 0001 in the 2's complement number system and confirm your results using the
            4-bit binary adder/subtractor in Figure 13-2. Also show the decimal equivalent of the
            number and your answers.

Questions: Did your calculated answer match the answer on the adder?

Was the decimal equivalent of your answer correct?

Was the sign bit correct?

Step 8.   Subtract 0001 from 0110 in the 2's complement number system and confirm your results
            using the 4-bit binary adder/subtractor in Figure 13-2. Also show the decimal equivalent of
            the numbers and your answer.

Questions: Did your calculated answer match the answer on the adder?

Was the decimal equivalent of your answer correct?

Was the sign bit correct?

Step 9.   Subtract 0110 from 0001 in the 2's complement number system and confirm your results
            using the 4-bit binary adder/subtractor in Figure 13-2. Also show the decimal equivalent of
            the numbers and your answer.

Questions: Did your calculated answer match the answer on the adder?

Was the decimal equivalent of your answer correct?

Was the sign bit correct?

Step 10.   Add 1101 to 0011 in the 2's complement number system and confirm your results using the
            4-bit binary adder/subtractor in Figure 13-2. Also show the decimal equivalent of the
            numbers and your answer.

134

Part II   Arithmetic Logic Circuits

Questions: Did your calculated answer match the answer on the adder?

Was the decimal equivalent of your answer correct?

Was the sign bit correct?

Step 11.   Subtract 0011 from 0011 in the 2's complement number system and confirm your results using the 4-bit binary adder/subtractor in Figure 13-2. Also show the decimal equivalent of the numbers and your answer.

Questions: Did your calculated answer match the answer on the adder?

Was the decimal equivalent of your answer correct?

Was the sign bit correct?

Step 12.   Subtract 1101 from 0011 in the 2's complement number system and confirm your results using the 4-bit binary adder/subtractor in Figure 13-2. Also show the decimal equivalent of the numbers and your answer.

Questions: Did your calculated answer match the answer on the adder?

Was the decimal equivalent of your answer correct?

Was the sign bit correct?

Step 13.   Subtract 0011 from 1101 in the 2's complement number system and confirm your results using the 4-bit binary adder/subtractor in Figure 13-2. Also show the decimal equivalent of the numbers and your answer.

Questions: Did your calculated answer match the answer on the adder?

Was the decimal equivalent of your answer correct?

Was the sign bit correct?

Step 14.   Convert decimal +6 and decimal –4 to 4-bit 2's complement binary numbers and add them in the space provided. Confirm your results using the 4-bit binary adder/subtractor in Figure 13-2.

Questions: Did your calculated answer match the answer on the adder?

Was the decimal equivalent of your answer correct?

Was the sign bit correct?

Step 15.   Convert decimal –2 and decimal +5 to 4-bit 2's complement binary numbers and add –2 to +5 in binary in the space provided. Confirm your results using the 4-bit binary adder/subtractor in Figure 13-2.

Questions: Did your calculated answer match the answer on the adder?

Was the decimal equivalent of your answer correct?

Was the sign bit correct?

EXPERIMENT

BCD Adder

Objectives:

1. Investigate the operation of a BCD adder.
2. Obtain experience adding BCD numbers and correcting the sum.

Materials:

One 5 V dc power supply
Eight logic switches
Thirteen logic probe lights
Two 4-bit binary parallel adders (2-7483 ICs)
One two-input AND gate (1-7408 IC)
Two two-input OR gates (1-7432 IC)

Theory:

In the **binary coded decimal (BCD)** number system, each decimal number is represented by a **4-bit binary code** from zero (0000) through nine (1001). The remaining six 4-bit numbers (1010-1111) are not part of the BCD code. Therefore, when BCD numbers are added, any answer that is larger than nine (1001) must be adjusted to the correct BCD code between zero (0000) and nine (1001), and a **carry** must be generated. The binary answer is corrected by adding **binary six (0110)** to the sum in order to skip the six invalid numbers. If the sum is nine (1001) or less, no adjustment is necessary.

A **BCD adder** is a 4-bit binary adder that adds two 4-bit binary numbers and determines if the sum is greater than nine (1001). If the sum is greater than nine (1001), binary six (0110) is added to the sum and a carry is generated. If the sum is not greater than nine (1001), binary zero (0000) is added and no carry is generated.

The BCD adder shown in Figure 14-1a includes two **4-bit binary adders** and a **logic network** that detects when the first adder output is greater than nine (1001). Each of the two 4-bit binary adders consists of a bank of four full-adders. The first 4-bit binary adder (bank of four full-adders) will add the two 4-bit BCD inputs. The second 4-bit binary adder (bank of four full-adders) will add **binary six (0110)** or **binary zero (0000)** to the sum, depending on the output from the first adder.

The **logic network** in Figure 14-1a will produce a binary one (1) output (X) when the output of the first adder is greater than nine (1001) or the first adder carry output (Co$'$) is binary one (1). Otherwise, the logic network will produce a logic zero (0) output (X). This logic network output (X) will apply a binary one (1) or zero (0) to the BCD adder carry output (Cout) and to the second and third B input bits of the second 4-bit binary adder. Because the first and fourth B input bits of the second 4-bit

137

binary adder are connected to ground (0), the second adder will add 0110 (binary six) or 0000 (binary zero) to the sum from the first adder, depending on the output (X) of the logic network.

If this experiment is performed in a **hardwired laboratory**, two **7483 4-bit binary adders** should be wired in place of the two banks of four full-adders, as shown in Figure 14-1b. The logic circuit inside each 7483 is similar to each bank of four full-adders in Figure 14-1a. Pin 1 on each 7483 is located in the lower left hand corner of the chip, as designated by the dot in Figure 14-1b. See Appendix A for the pin diagram of the 7483 IC chip.

To add two **2-digit decimal numbers** using the BCD code, two BCD adders must be cascaded. The **carry output (Cout)** of the first BCD adder must be connected to the **carry input (Cin)** of the second BCD adder. The number of BCD adders required to add numbers in the BCD code is determined by the number of digits in the decimal numbers to be added.

Figure 14-1a BCD Adder

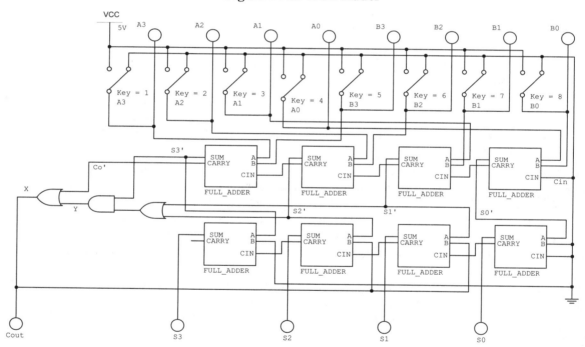

Figure 14-1b 7483 BCD Adder

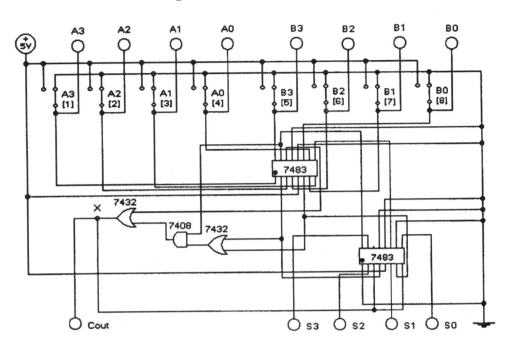

Procedure:

Step 1.   Open circuit file FIG14-1. You are looking at a BCD adder constructed from two banks of four full-adders. See the Theory section for more details. You will use this BCD adder to confirm the results when two BCD numbers are added. Press number keys 1-8 to control the logic switches. If you are performing this experiment in a **hardwired laboratory**, wire two 7483 4-bit binary adders as shown in Figure 14-1b.

Step 2.   Add BCD 0111 to BCD 0010 in the space provided and confirm your results using the BCD adder in Figure 14-1. Also show the decimal equivalent of the numbers and your answer.

Questions: Did your calculated answer match the answer on the adder?

Was the decimal equivalent of your answer correct?

Step 3.   Add BCD 1001 to BCD 0110 in the space provided and confirm your results using the BCD adder in Figure 14-1. Also show the decimal equivalent of the numbers and your answer.

Questions: Did your calculated answer match the answer on the adder?

Was the decimal equivalent of your answer correct?

Step 4.   Add BCD 1001 to BCD 1000 in the space provided and confirm your results using the BCD adder in Figure 14-1. Also show the decimal equivalent of the numbers and your answer.

Questions: Did your calculated answer match the answer on the adder?

Was the decimal equivalent of your answer correct?

Step 5.   Convert decimal 7 and decimal 5 to BCD numbers and add them in the space provided. Confirm your results using the BCD adder in Figure 14-1.

Questions: Did your calculated answer match the answer on the adder?

Was the decimal equivalent of your answer correct?

Step 6.   Use a 4-input truth table with inputs S3′, S2′, S1′, and S0′ to prove that the logic network in Figure 14-1a will produce a binary one (1) at the output of the 7408 AND gate (Y) when the binary output of the first adder is between ten (1010) and fifteen (1111).

EXPERIMENT

Parity Generator/Checker

Objectives:

1. Demonstrate how XOR gates are used to build a parity checker circuit.
2. Demonstrate how XOR gates are used to build a parity generator circuit.
3. Demonstrate how the 74280 parity generator/checker is used to build a parity error-detection system.
4. Demonstrate the difference between an even and an odd parity system.

Materials:

One 5 V dc voltage supply
Five logic switches
Eleven logic probe lights
One INVERTER (1-7404 IC)
Three XOR gates (1-7486 IC)
Two parity generator/checkers (2-74280 ICs)
Four 1 kΩ resistors

Theory:

When a **digital code** is being transmitted from one point to another, **line noise** can introduce an **error** in one of the bits. This may cause a binary one to be received as a binary zero or a binary zero to be received as a binary one. One of the methods used to detect this type of error is the **parity method of error detection.** In this method of error detection, a **parity bit** is added to each transmitted binary code to make the total number of binary ones **even or odd** (depending on the system). If the number of binary ones received is not the same parity (even or odd) as the parity of the transmitted code, there is a high probability that a transmission error has occurred. When the parity method of error detection is used, the parity (even or odd) must be the same at both the transmitter and the receiver so that the parity checker at the receiver knows whether to check for even or odd parity.

If the system is an **even parity system**, the parity generator will generate a binary one parity bit when the transmitted code has an odd number of binary ones. This will make the transmitted code have an even number of binary ones. The parity generator will generate a binary zero parity bit when the transmitted code has an even number of binary ones, keeping the number of transmitted binary ones even. If the system is an **odd parity system**, the parity generator will generate a binary one parity bit when the transmitted code is even parity, and a binary zero parity bit will be generated when the transmitted code is odd parity.

If there is a change in more than one bit in any of the binary codes transmitted, then an error may not be detected. Therefore, the parity method of error detection is not perfect, but it has a high probability of success because there is a low probability that two bits will change in a single transmitted code. More reliable methods of **error detection** are available, but they are more costly to implement.

A **parity checker** can be constructed using XOR gates because an XOR gate produces a binary one output when there is an odd number of binary ones on the input terminals. A 4-bit parity checker using three XOR gates is shown in Figure 15-1. The XOR gate output is used to detect an odd number of binary ones at the input. The inverted XOR gate output is used to detect an even number of binary ones at the input.

A **parity generator** can be constructed using XOR gates, as shown in Figure 15-2. When there is an odd number of binary ones input to the parity generator, the parity generator circuit (XOR gates and INVERTER) will produce a binary zero parity bit. This will cause an odd number of binary ones to be transmitted (odd parity). When there is an even number of binary ones input to the parity generator, the parity generator circuit will produce a binary one parity bit, causing an odd number of binary ones to be transmitted again (odd parity). This means that the parity generator circuit in Figure 15-2 is an **odd parity generator**.

The **74280 parity generator/checker**, shown in Figure 15-3, can be used as a parity generator or a parity checker. When there is an odd number of binary ones on the 74280 inputs (A-I), the odd output will go to a binary one. When there is an even number of binary ones on the 74280 inputs (A-I), the even output will go to a binary one. When the 74280 is used as a parity generator, the even or odd output is used to generate the parity bit, depending on whether the system is even or odd parity. For an even parity system, the odd parity output is used. This will make the total number of binary ones transmitted always be even. For an odd parity system, the even parity output is used. This will make the total number of binary ones transmitted always be odd. The circuit in Figure 15-3 is using only four of the 74280 inputs. Therefore, the other four inputs have been connected to ground (binary zero input) on both the parity generator and the parity checker. The 7486 XOR gate in Figure 15-3 is being used as a controlled inverter. When the INV key is down (0), the transmitted A bit is passed to the parity checker unchanged. When the INV key is up (1), the transmitted A bit is inverted, causing the received parity to be different (odd parity) than the transmitted parity (even parity). This will cause an error signal at the output of the parity checker.

Figure 15-1 Four-Bit Parity Checker

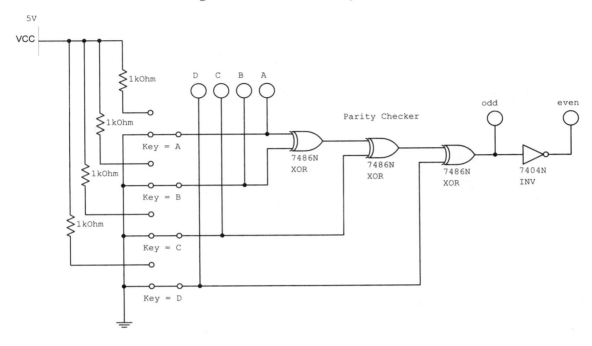

Figure 15-2 Four-Bit Parity Generator

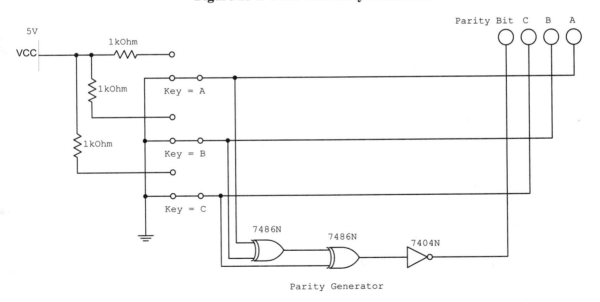

Figure 15-3 Parity Error-Detection System

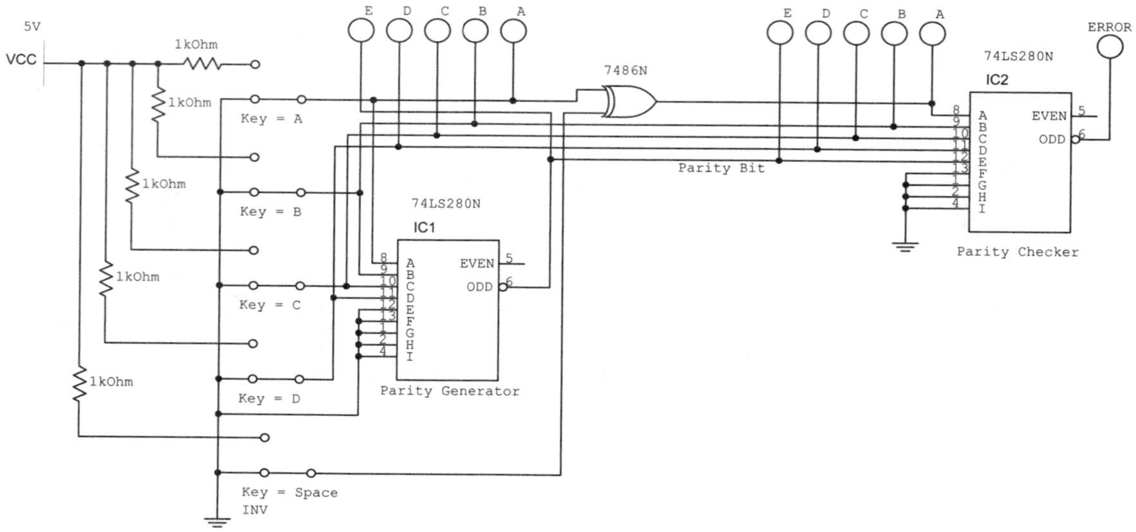

Procedure:

Step 1.   Open circuit file FIG 15-1. For the circuit in Figure 15-1, record the **predicted** odd and even output logic levels for each input combination in Table 15-1.

Table 15-1

				Predicted		Measured	
D	C	B	A	odd	even	odd	even
0	0	0	0				
0	0	0	1				
0	0	1	0				
0	0	1	1				
0	1	0	0				
0	1	0	1				
0	1	1	0				
0	1	1	1				
1	0	0	0				
1	0	0	1				
1	0	1	0				
1	0	1	1				
1	1	0	0				
1	1	0	1				
1	1	1	0				
1	1	1	1				

Step 2.   Click the On-Off switch to run the simulation. By switching the logic switches to the appropriate positions, record the **measured** odd and even output logic levels for each input combination in Table 15-1. You can switch the logic switches by pressing the letter on the computer keyboard corresponding to the letter label on the switch. After completing the table, stop the simulation.

Questions: How did your predicted outputs compare with the measured values?

What is the logic circuit in Figure 15-1 doing?

Step 3.   Open circuit file FIG 15-2. For the circuit in Figure 15-2, record the **predicted** parity bit
               output logic level for each input combination in Table 15-2.

Table 15-2

C	B	A	Predicted Parity Bit	Measured Parity Bit
0	0	0		
0	0	1		
0	1	0		
0	1	1		
1	0	0		
1	0	1		
1	1	0		
1	1	1		

Step 4.   Click the On-Off switch to run the simulation. By switching the logic switches to the
               appropriate positions, record the **measured** parity bit output logic levels for each input
               combination in Table 15-2. You can switch the logic switches by pressing the letter on the
               computer keyboard corresponding to the letter label on the switch. After completing the
               table, stop the simulation.

Question: How did your predicted outputs compare with the measured values?

Is this circuit an even parity generator or an odd parity generator?

Step 5.   Open circuit file FIG15-3. Notice that a 4-bit binary code is generated by the settings of
               switches A, B, C, and D, with a fifth bit (parity bit E) generated by the parity generator
               (IC1). This 5-bit code is being transmitted over a 5-bit transmission line to a 5-bit parity
               checker (IC2) on the other end of the line. Also notice that a controlled inverter (7486) has
               been inserted in one of the transmission lines (line A) to make it possible to simulate a bit
               error. When the INV switch is down (0), the binary bit on line A is passed through the
               controlled inverter unchanged. When the INV switch is up (1), the binary bit on line A is
               inverted, causing the received code input to the parity checker to have a different parity
               than the transmitted code. This will simulate an error in the data transmission, which should
               cause the error light on the parity checker to light. Review the Theory section of this
               experiment for further discussion of the 74280 parity generator/checker.

Step 6.   Click the On-Off switch to run the simulation. By pressing the A, B, C, and D keys on the computer keyboard, test each 4-bit input combination to generate the various binary codes, then stop the simulation.

Questions: Were the received binary codes the same as the transmitted binary codes?

Did the error light turn on for any of the transmitted codes? **Explain the results.**

Step 7.   Start the simulation again and press the space bar on the keyboard to switch the INV switch to the up (1) position. By pressing the A, B, C, and D keys on the computer keyboard, test each 4-bit input combination to generate the various codes again, then stop the simulation.

Questions: Were the received binary codes the same as the transmitted binary codes?

Did the error light turn on for any of the transmitted codes? **Explain the results.**

Is this an even or an odd parity error-detection system? **Explain your answer.**

EXPERIMENT

Magnitude Comparator

Objectives:

1. Demonstrate how the XNOR gate (XOR gate and INVERTER) is used to build a magnitude comparator that detects two equal 2-bit binary numbers.
2. Demonstrate how the XNOR gate is used to build a magnitude comparator that detects two equal 4-bit binary numbers.

Materials:

One 5 V dc voltage supply
Eight logic switches
Nine logic probe lights
Four INVERTERS (1-7404 IC) *SN 74LS04 N*
Four XOR gates (1-7486 IC) *SN 74LS86N*
One two-input AND gate (1-7408 IC) *HD 74LS08P*
One four-input AND gate (1-7421 IC)
Four 1 kΩ resistors

Theory:

The function of a digital **magnitude comparator** is to compare the magnitudes of two binary numbers to determine their relationship. In its basic form, a magnitude comparator determines whether two binary numbers are equal. Most **integrated circuit magnitude comparators**, such as the 7485, have **three outputs**. One output produces a binary one when the binary numbers are **equal**, the second output produces a binary one when the **A binary number is greater than the B number**, and the third output produces a binary one when the **B binary number is greater than the A number**. In this experiment, you will demonstrate how XNOR gates can be used to build a comparator that will produce a binary one output when two binary numbers are equal in magnitude. Review the discussion of the XNOR gate in the Theory section of Experiment 11 before proceeding with this experiment.

Because the XNOR gate (XOR with an INVERTER on the output) produces a binary one output when both inputs are equal, it is an ideal gate for building a magnitude comparator. The circuit in Figure 16-1 will be used to demonstrate a **2-bit magnitude comparator** using XNOR gates. This circuit will compare the 2-bit binary numbers A1 A0 and B1 B0 and produce a binary one output when both numbers are equal. Notice that the **two least-significant bits** of the 2-bit binary numbers (A0 and B0) are applied to the inputs of one XNOR gate and the **two most-significant bits** of the 2-bit binary numbers (A1 and B1) are applied to the inputs of the other XNOR gate. When the two binary numbers are equal, their corresponding bits will be equal, causing both XNOR gates to produce a binary one output to both AND gate inputs. This will cause the AND gate to produce a binary one output, lighting the output logic probe light (A = B).

The circuit in Figure 16-2 will be used to demonstrate a **4-bit magnitude comparator** using XNOR gates. This circuit will compare two 4-bit binary numbers (A3 A2 A1 A0 and B3 B2 B1 B0) and produce a binary one output when both 4-bit numbers are equal. The theory for this 4-bit magnitude comparator is the same as the theory for the 2-bit magnitude comparator, described previously. Notice that bits A0 and B0 are applied to the first XNOR gate, bits A1 and B1 are applied to the second XNOR gate, bits A2 and B2 are applied to the third XNOR gate, and bits A3 and B3 and applied to the fourth XNOR gate. When all of the corresponding bits are equal, four binary ones (1) will be applied to the four AND gate inputs. This will produce a binary one (1) at the AND gate output (A = B).

Figure 16-1 Two-Bit Magnitude Comparator

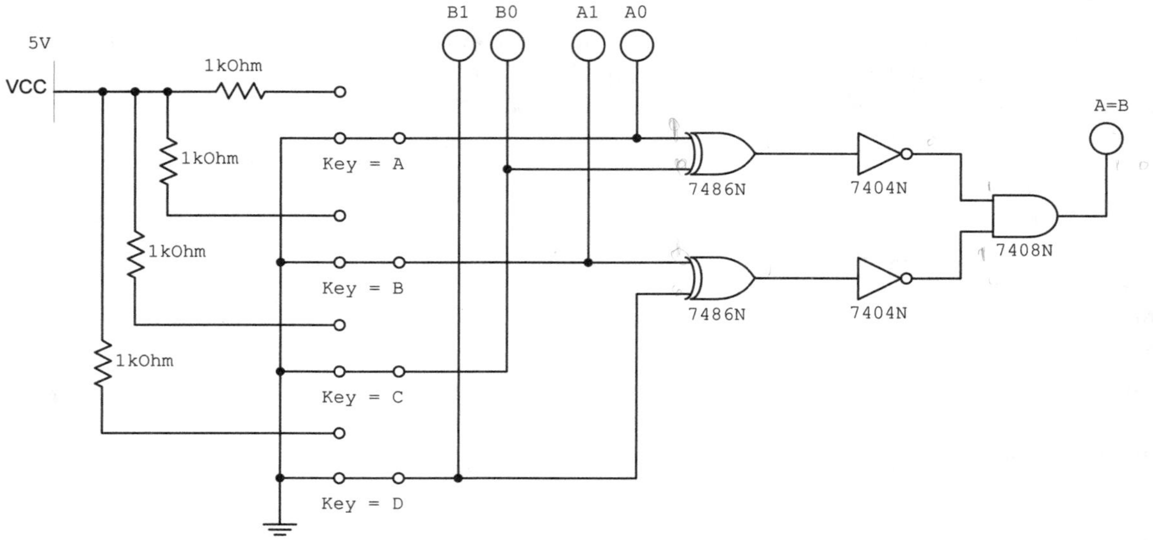

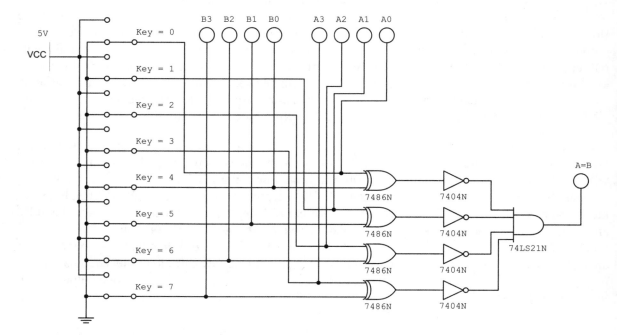

Figure 16-2 Four-Bit Magnitude Comparator

Procedure:

Step 1.   Open circuit file FIG16-1. For the circuit in Figure 16-1, record the **predicted** output
(A = B) for each input combination in Table 16-1.

Table 16-1

B1	B0	A1	A0	Predicted A = B	Measured A = B
0	0	0	0	1	1
0	0	0	1	0	0
0	0	1	0	0	0
0	0	1	1	0	0
0	1	0	0	0	0
0	1	0	1	1	1
0	1	1	0	0	0
0	1	1	1	0	0
1	0	0	0	0	0
1	0	0	1	0	0
1	0	1	0	1	1
1	0	1	1	0	0
1	1	0	0	0	0
1	1	0	1	0	0
1	1	1	0	0	0
1	1	1	1	1	1

Step 2.   Click the On-Off switch to run the simulation. By switching the logic switches to the
appropriate positions, record the **measured** output (A = B) logic levels for each input
combination in Table 16-1. You can switch the logic switches by pressing the letter on the
computer keyboard corresponding to the letter label on the switch. After the table is
completed, stop the simulation.

Questions: How did your predicted outputs compare with the measured values?

CORRECTLY

What is the logic circuit in Figure 16-1 doing?

COMPARING THE BINARY WORDS AND LIGHTING UP
WHEN THEY MATCH.

Step 3.   Open circuit file FIG16-2. For the circuit in Figure 16-2, record the **predicted** output
(A = B) for each input combination in Table 16-2.

Table 16-2

B3	B2	B1	B0	A3	A2	A1	A0	Predicted A = B	Measured A = B
0	0	0	0	0	0	0	0	1	1
0	0	0	1	0	0	0	1	1	1
0	0	1	0	0	0	1	0	1	1
0	0	1	1	0	0	1	1	1	1
0	1	0	0	0	1	0	0	1	1
0	1	0	1	0	1	0	1	1	1
0	1	1	0	0	1	1	0	1	1
0	1	1	1	0	1	1	1	1	1
1	0	0	0	0	0	0	0	0	0
1	0	0	1	0	0	0	1	0	0
1	0	1	0	0	0	1	0	0	0
1	0	1	1	0	0	1	1	0	0
1	1	0	0	0	1	0	0	0	0
1	1	0	1	0	1	0	1	0	0
1	1	1	0	0	1	1	0	0	0
1	1	1	1	0	1	1	1	0	0

Step 4.   Click the On-Off switch to run the simulation. By switching the logic switches to the
appropriate positions, record the **measured** output (A = B) logic levels for each input
combination in Table 16-2. You can switch the logic switches by pressing the number on
the computer keyboard corresponding to the label on the switch. After the table is
completed, stop the simulation.

Questions: How did your predicted outputs compare with the measured values?

CORRECTLY

What is the logic circuit in Figure 16-2 doing?

COMPARING BOTH WORDS AND LIGHTING UP WHEN THEY MATCH

EXPERIMENT

17 Troubleshooting Arithmetic Circuits

Objectives:

1. Determine the defective logic gate or component for various arithmetic logic circuits by monitoring the logic levels at circuit test points.

Materials:

This experiment can only be performed on Electronics Workbench Multisim using the circuits disk provided with this manual.

Theory:

In order to perform this experiment effectively, you must first complete Experiments 11–16. Use the theory learned in those experiments to find the defective logic gate or component in the arithmetic circuits in this experiment.

Determine the defective component in each experiment by changing the logic circuit binary inputs until you find the binary inputs that cause the circuit output to be incorrect. Observe the inputs and outputs of each logic gate when the circuit output is incorrect. A logic gate is defective if it has an incorrect output for any combination of binary inputs. Remember that an open input on a 7400 series logic gate will behave as if there is a logical "one" on that open input terminal, even if that open input is being caused by the open output of another logic gate or the logic probe light is indicating a zero. For example, if one of the inputs of an OR gate that is not defective is connected to a logic gate with an open output, the OR gate will act as if there is a logical "one" on that input and a logical "one" on its output (logic probe light "on"). This will happen because an OR gate produces a logical "one" at the output when any input is at a logical "one" (or open). For this reason, you may need to try all possible logic network binary input combinations before concluding which logic gate is defective.

Procedure:

Don't forget to read the Theory section before attempting to determine the defective components in the following arithmetic circuits.

1.   Open circuit file FIG17-1. Click the On-Off switch to run the simulation. Based on the logic levels at the full-adder test points for different logic inputs, determine which logic gate is defective. To switch the logic switches, press the key on the computer keyboard that matches the label on the switch.

         Defective gate_____

2.   Open circuit file FIG17-2. Click the On-Off switch to run the simulation. Based on the logic levels at the full-adder test points for different logic inputs, determine which logic gate is defective. To switch the logic switches, press the key on the computer keyboard that matches the label on the switch.

         Defective gate_____

3.   Open circuit file FIG17-3. Click the On-Off switch to run the simulation. Based on the logic levels at the full-adder test points for different logic inputs, determine which logic gate is defective. To switch the logic switches, press the key on the computer keyboard that matches the label on the switch.

         Defective gate_____

4.   Open circuit file FIG17-4. Click the On-Off switch to run the simulation. Based on the logic levels at the full-adder test points for different logic inputs, determine which logic gate is defective. To switch the logic switches, press the key on the computer keyboard that matches the label on the switch.

         Defective gate_____

5.   Open circuit file FIG17-5. Click the On-Off switch to run the simulation. Based on the logic levels at the full-adder test points for different logic inputs, determine which logic gate is defective. To switch the logic switches, press the key on the computer keyboard that matches the label on the switch.

         Defective gate_____

6.   Open circuit file FIG17-6. Click the On-Off switch to run the simulation. Using the logic probe light to measure the logic levels at various test points in the full-adder circuit for different logic inputs, determine which logic gate is defective. To switch the logic switches, press the key on the computer keyboard that matches the label on the switch.

         Defective gate_____

7.   Open circuit file FIG17-7. Click the On-Off switch to run the simulation. Using the logic probe light to measure the logic levels at various test points in the full-adder circuit for different logic inputs, determine which logic gate is defective. To switch the logic switches, press the key on the computer keyboard that matches the label on the switch.

        Defective gate_____

8.   Open circuit file FIG17-8. Click the On-Off switch to run the simulation. Using the logic probe light to measure the logic levels at various test points in the full-adder circuit for different logic inputs, determine which logic gate is defective. To switch the logic switches, press the key on the computer keyboard that matches the label on the switch.

        Defective gate_____

9.   Open circuit file FIG17-9. Click the On-Off switch to run the simulation. Using the logic probe light to measure the logic levels at various test points in the magnitude comparator circuit for different logic inputs, determine which logic gate is defective. To switch the logic switches, press the key on the computer keyboard that matches the label on the switch.

        Defective gate_____

10.   Open circuit file FIG17-10. Click the On-Off switch to run the simulation. Using the logic probe light to measure the logic levels at various test points in the adder/subtractor circuit for different logic inputs, determine which logic gate is defective. To switch the logic switches, press the key on the computer keyboard that matches the label on the switch.

        Defective gate_____

III

MSI Logic Circuits

The experiments in Part III involve the study of **medium scale integrated (MSI) logic circuits** such as **decoders, encoders, multiplexers**, and **demultiplexers**. In the final experiment in Part III you will solve some **troubleshooting problems** involving these devices.

The circuits for the experiments in Part III can be found on the enclosed disk in the PART3 subdirectory.

EXPERIMENT

18 | Decoders and Encoders

Objectives:

1. Investigate the operation of a 74138 3-line-to-8-line decoder.
2. Demonstrate how to wire a 4-line-to-16-line decoder using two 74138 3-line-to-8-line decoders.
3. Investigate the operation of a 7442 BCD-to-decimal decoder.
4. Investigate the operation of a BCD-to-7-segment decoder/driver driving a 7-segment LED display.
5. Investigate the operation of a 74147 decimal-to-BCD priority encoder.
6. Demonstrate an encoder-decoder circuit using an 8-line-to-3-line encoder and a BCD-to-7-segment decoder/driver.
7. Design a 2-line-to-4-line decoder using AND gates and inverters.

Materials:

One 5 V dc power supply
Ten logic switches
Twenty logic probe lights
Two 3-line-to-8-line decoders (2-74138 ICs)
One BCD-to-decimal decoder (1-7442 IC)
One BCD-to-7-segment decoder/driver (1-7447 IC)
One common anode 7-segment LED display
One decimal-to-BCD encoder (1-74147 IC)
Four INVERTERS (1-7404 IC)
One 8-line-to-3-line encoder (1-74148 IC)
Four two-input AND gates (1-7408 IC)
Seven 300 Ω resistors

Theory:

A decoder is a logic circuit that accepts a binary input code and activates only the output or outputs that corresponds to the binary input code. Some decoders have one or more **enable inputs** that are used to enable or disable the decoder.

The 74138 in Figure 18-1 is a **3-line-to-8-line octal decoder** capable of decoding a 3-bit binary input (000-111) into one of eight separate **active low** outputs. It also has three enable inputs (G1, G2A, and G2B). **Enable input G1** must be **high (1)** and **enable inputs G2A and G2B** must be **low (0)** to enable the 74138 decoder. When the decoder is **disabled**, all outputs are at a **logical high (1)**. When the decoder is **enabled**, the binary code on **inputs A, B, and C** determines which output will go **low (0)**.

The circuit in Figure 18-2 will demonstrate the **timing** of the 74138 decoder **output waveforms** using a **logic analyzer** and a **word generator**. The word generator will transmit binary numbers from 000 to 111 in a two cycle sequence while the logic analyzer monitors the eight output terminals and the 3-bit input. If this part of the experiment is performed in a hardwired laboratory, a **3-bit binary counter** can be used in place of the word generator.

The circuit in Figure 18-3 shows how to connect two 74138 3-line-to-8-line decoders to make a **4-line-to-16-line (1-of-16) decoder**. The fourth input (D) is connected to enable terminal G1 on the second 74138 decoder and enable terminal G2A (inverted input) on the first 74138 decoder. Therefore, the first decoder will display outputs for binary inputs 0000 through 0111 (0–7) when D is binary zero (0), placing a binary zero on enable G2A of the first decoder. The second decoder will display outputs for binary inputs 1000 through 1111 (8–15) when D is binary one (1), placing a binary one (1) on enable G1 of the second decoder.

The 7442 in Figure 18-4 is a **BCD-to-decimal (4-line-to-10-line) decoder** with 10 **active low outputs** (0–9). When a binary BCD input (DCBA) between zero (0000) and nine (1001) is applied, the appropriate output will go low (0). When an input (DCBA) outside the range of zero (0000) to nine (1001) is applied, all of the outputs remain high (1) because the BCD code only includes binary numbers between zero and nine. The 7442 does not have an enable input. Therefore, one of the outputs will go low (0) as soon as a binary input between zero and nine is applied.

The 7447 in Figure 18-5 is a **BCD-to-7-segment decoder/driver** connected to a **7-segment LED display**. The LEDs will display the decimal equivalent of the BCD code applied to the decoder/driver input (DCBA) for numbers between zero and nine. Numbers outside the range of zero to nine do not exist in the BCD code. Because the 7447 has **active low outputs**, a **common anode LED display** is required. **External resistors** are needed because they are not built into the decoder/driver circuit and the diode currents need to be limited. If this part of the experiment is performed in a hardwired laboratory, a 7447 decoder/driver should be used with a common anode 7-segment LED display.

Encoding is the opposite process from decoding. An **encoder** generates a **binary coded output** from a **singular active input**. Some encoders have an **enable input** that may be used to enable or disable the encoder. Some encoders also have an output that is active only when none of the inputs are active. This output can be used to control external logic circuitry or blank an LED display.

The 74147 in Figure 18-6 is a **decimal-to-BCD priority encoder** with **active low inputs** and **active low outputs**. A **priority encoder** will only respond to the input that represents the value with the **highest priority** (usually the highest value) when more than one input is active. Because of the active low outputs, INVERTERS are required for each output terminal. Without the INVERTERS, the encoder would produce an **inverted BCD output code**. The 74147 has only nine inputs (input zero is missing). **Input zero (0) is not connected** to the 74147 encoder because the encoder assumes a decimal zero (0) input when none of the inputs are active and will output a BCD zero (0000) code.

The circuit in Figure 18-7 will demonstrate how an **encoder**, a **7-segment decoder/driver**, and an **LED display** can be used to display the decimal equivalent of a key closure. A **74148 8-line-to-3-line priority encoder** is used because it has an output (E0) that will go high (1) when one of the inputs is active and stays low (0) when none of the inputs are active. This output (E0) is used to blank the 7-segment display until one of the encoder inputs is active. This is accomplished by connecting the encoder E0 output to the decoder/driver **active low blanking input (BI/RBO)**. When none of the encoder inputs are active, the encoder E0 output will bring the decoder/driver active low blanking input (BI/RBO) to binary zero (0), blanking the 7-segment display. When one of the encoder inputs is active, encoder output E0 will bring the decoder/driver blanking input (BI/RBO) to binary one (1), allowing the decimal output to be displayed on the 7-segment LED. Because the **74148 encoder** has only **three outputs**, input D of the decoder/driver is connected to ground (0). Therefore, only decimal numbers 0–7 can be displayed. The 74148 encoder also has an **active low enable input (EI).** Therefore, this input (EI) must be connected to ground (0) to enable the 74148. Similar to the 74147, the 74148 has **active low inputs and outputs**. Therefore, INVERTERS are required on the 74148 encoder outputs also. When more than one of the inputs are active (low), the 74148 encoder output is controlled by the input that represents the highest decimal value because the 74148 is a **priority encoder**.

Figure 18-1 3-Line-to-8-Line (1-of-8) Decoder

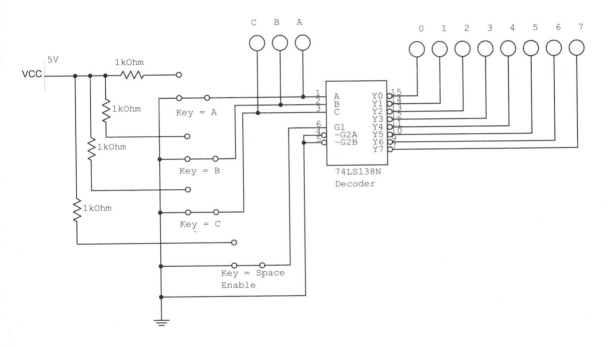

Figure 18-2 Decoder Output Waveform Analysis

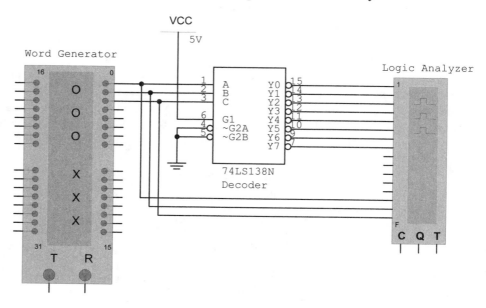

Figure 18-3   4-Line-to-16-Line (1-of-16) Decoder

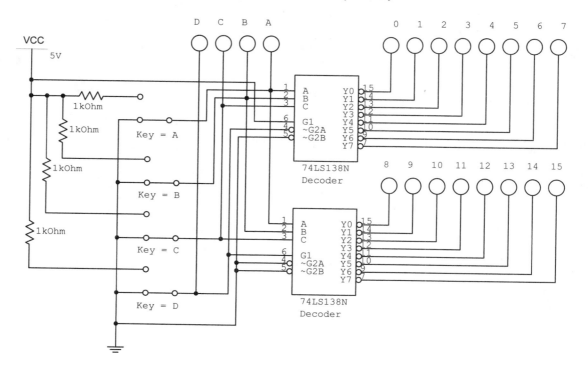

Figure 18-4 BCD-to-Decimal Decoder

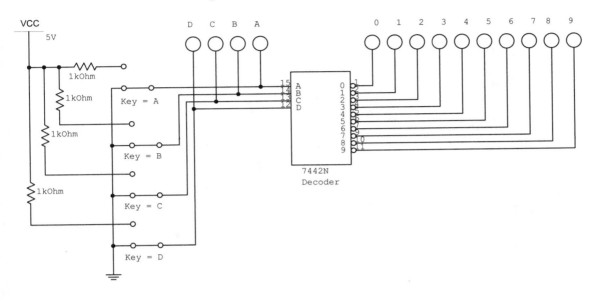

Figure 18-5 BCD-to-7-Segment Decoder/Driver

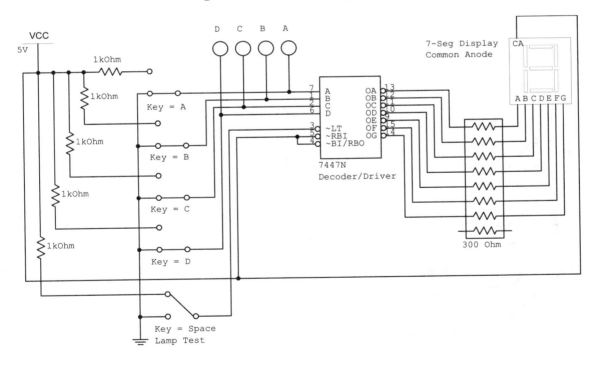

Figure 18-6 Decimal-to-BCD Priority Encoder

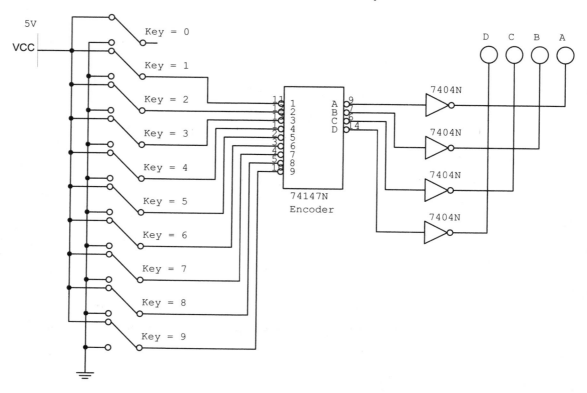

Figure 18-7 Encoder/Decoder Circuit

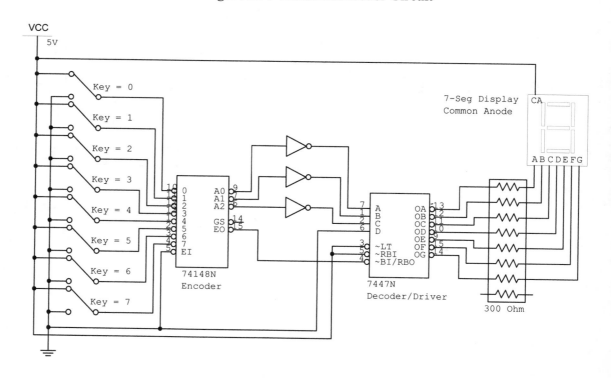

Procedure:

Step 1:   Open circuit file FIG18-1. You are looking at a 74138 3-line-to-8-line decoder circuit. All logic switches should be down. Click the On-Off switch to run the simulation. Notice that enable terminal G1 on the 74138 decoder is grounded (0).

Question: What do you notice about decoder outputs Y0-Y7 based on logic probe lights 0–7? Is the decoder enabled or disabled? **Explain why.**

DISABLED, BECAUSE THE ENABLE KEY IS SET TO 0.

THE PROBE LIGHTS ARE ACTIVE LOW

Step 2.   Set the ENABLE switch to binary one (up) by pressing the space bar on the computer keyboard. This places a binary one (1) on the 74138 decoder enable terminal G1. Notice that enable terminals G2A and G2B on the decoder are grounded (0). Also notice that the binary input (CBA) is 000.

Question: What do you notice about the decoder outputs Y0-Y7 based on logic probe lights 0–7? **Explain your answer.**

THE Y0 IS OFF, ALL OTHERS ARE ON

Step 3.   By pressing the A, B, and C keys on the keyboard, change the binary input to the values in Table 18-1 and record the outputs in the table. After the table is complete, stop the simulation.

Table 18-1 74138 Decoder

Inputs			Outputs							
C	B	A	Y0	Y1	Y2	Y3	Y4	Y5	Y6	Y7
0	0	0	0	1	1	1	1	1	1	1
0	0	1	1	0	1	1	1	1	1	1
0	1	0	1	1	0	1	1	1	1	1
0	1	1	1	1	1	0	1	1	1	1
1	0	0	1	1	1	1	0	1	1	1
1	0	1	1	1	1	1	1	0	1	1
1	1	0	1	1	1	1	1	1	0	1
1	1	1	1	1	1	1	1	1	1	0

Question: What conclusion con you draw about the relationship between the decoder outputs and the binary inputs? Are the outputs active low or active high?

Active low

Step 4.   Open circuit file FIG18-2. Notice that the decoder enable terminal G1 is connected to 5 V (1) and enable terminals G2A and G2B are grounded (0) to enable the decoder. Bring down the word generator enlargement and make sure that the following settings are selected: Frequency = 1 kHz, Trigger = Internal, Hex codes (00000000, 00000001, 00000002, 00000003, 00000004, 00000005, 00000006, 00000007, repeated twice). Move the word generator to the left by clicking and dragging. Bring down the logic analyzer enlargement and make sure that the following settings are selected: Clocks/Div = 16, Clock Setup (Clock Source = Internal, Clock Rate = 10 kHz, Pre-trigger Samples = 100, Post-trigger Samples = 1000, Threshold Voltage = 2.5), Trigger Settings (Trigger Clock Edge = Positive, Trigger Qualifier = x, Pattern A= xxxxxxxxxxxxxxxx, Trigger Combinations = A). Click BURST on the word generator to run the simulation, then stop the simulation after the curve plot is complete. Notice that the decoder outputs (Y0–Y7) and the decoder inputs (CBA) are being monitored by the logic analyzer. (The top eight curve plots are the decoder outputs and the bottom three curve plots are the decoder inputs). Also notice that the word generator transmitted binary 000 (0) through binary 111 (7) to the decoder input (CBA) in a two cycle sequence in the burst mode.

NOTE: If this part of the experiment is performed in a hardwired laboratory, a 3-bit counter can be used in place of the word generator.

Question: What is the relationship between the decoder outputs (Y0-Y7) and the binary inputs (CBA) generated by the word generator?

Step 5.   Open circuit file FIG18-3. You are looking at two 74138 3-line-to-8-line decoders wired as a 4-line-to-16-line decoder. Notice that the fourth input (input D) is connected to enable G1 on the bottom 74138 decoder and enable G2A (inverted input) on the top 74138 decoder. Also notice that inputs A, B, and C are connected to terminals A, B, and C on both 74138 decoders. Click the On-Off switch to run the simulation.

Question: What do the outputs indicate? **Explain.**

Step 6.   By pressing the A, B, C, and D keys on the keyboard, change the binary inputs to all possible input combinations and notice the outputs, then stop the simulation.

Questions: What conclusion can you draw about the relationship between the decoder outputs and the binary inputs? **Explain.**

Step 7.   Open circuit file FIG18-4. You are looking at a 7442 BCD-to-decimal (4-line-to-10-line) decoder circuit. Notice that the 7442 does not have an enable input for enabling or disabling the decoder. Click the On-Off switch to run the simulation. By pressing the A, B, C, and D keys on the keyboard, change the binary input to the values in Table 18-2 and record the outputs in the table, then stop the simulation.

Table 18-2 7442 Decoder

Inputs				Outputs									
D	C	B	A	0	1	2	3	4	5	6	7	8	9
0	0	0	0										
0	0	0	1										
0	0	1	0										
0	0	1	1										
0	1	0	0										
0	1	0	1										
0	1	1	0										
0	1	1	1										
1	0	0	0										
1	0	0	1										
1	0	1	0										
1	0	1	1										
1	1	0	0										
1	1	0	1										
1	1	1	0										
1	1	1	1										

Questions: What conclusion can you draw about the relationship between the decoder outputs and the binary inputs? Are the outputs active low or active high?

What happens when a binary input outside the range of 0–9 is applied?

Step 8.   Open circuit file FIG18-5. You are looking at a BCD-to-7-segment decoder/driver connected to a 7-segment LED display. Because the decoder/driver has active low outputs, the LED display is a common anode display. This decoder circuit will display the decimal equivalent of the BCD code applied to the decoder/driver input. Click the On-Off switch to run the simulation. Press the space bar on the keyboard to switch the Lamp Test switch to ground (down). This will place a binary zero (0) on the active low lamp test (LT) terminal on the decoder/driver, which will cause all of the LED segments to light on the 7-segment display if the display is functioning properly.

NOTE: If this part of the experiment is performed in a hardwired laboratory, a 7447 decoder/driver should be used with a common anode 7-segment LED display.

Question: Did the lamp test verify that your 7-segment LED display is working properly?

Step 9.   Press the space bar to place a binary one on the decoder active low lamp test (LT) terminal and remove the lamp test. By pressing the A, B, C, and D keys on the keyboard, try different binary inputs to verify that the LED display shows the correct decimal equivalents, then stop the simulation.

Questions: Does your LED display show the correct decimal numbers for the binary inputs 0000 through 1001?

What happens when a binary number outside the range of zero (0000) through nine (1001) is applied to the decoder input?

Step 10.   Open circuit file FIG18-6. You are looking at a decimal-to-BCD priority encoder circuit. Click the On-Off switch to run the simulation. Notice what happens to the binary output (DCBA) when you open and close the numbered switches connected to the 74147 encoder inputs. You can open and close the numbered switches by pressing the equivalent number keys on the computer keyboard. After you have tried all of the input keys, stop the simulation.

Questions: What is the relationship between the numbered switches and the binary output (DCBA)?

Are the inputs on the 74147 encoder active low or active high?

What happens when two input number switches are low at the same time? Which input controls the BCD output? **Explain why.**

Why is the input zero (0) switch not connected to the encoder? What is the 74147 encoder BCD output when none of the inputs are low?

Are the outputs on the 74147 encoder active low or active high? What is the purpose of the four INVERTERS connected to the encoder outputs?

Step 11.   Open circuit file FIG18-7. You are looking at a 74148 8-line-to-3-line priority encoder feeding a BCD-to-7-segment decoder and LED display. Because the decoder has four inputs (DCBA) and the encoder has three outputs, input D of the decoder is connected to ground (0). Therefore, the decoder will only count to seven (binary 111). Also, output E0 of the 74148 encoder produces a binary zero (0) output when none of the encoder inputs are active. Because E0 is connected to the active low blanking input (BI/RBO) of the decoder, the decoder output will be blanked-out when none of the encoder inputs are active. Therefore, the LED display will only display a decimal number when one of the encoder input number keys is active (low). Click the On-Off switch to run the simulation. Activate the number switches by pressing the equivalent number keys on the keyboard and notice the readings on the LED display, then stop the simulation.

Questions: What is the relationship between the numbered switches and the decimal display?

Are the inputs on the 74148 encoder active low or active high?

What happens when two input number switches are active at the same time? Which input controls the decimal output? **Explain why.**

What happens when none of the number switches are active (0)? **Explain why.**

Are the outputs on the 74148 encoder active low or active high? What is the purpose of the three INVERTERS connected to the encoder outputs?

Is the enable (EI) terminal on the 74148 encoder active low or active high? Was the 74148 enabled?

Step 12.   Based on the truth table in Table 18-3, design a 2-line-to-4-line decoder with active high outputs using AND gates and INVERTERS. Design a separate logic circuit for each output, all connected to the same two inputs (B and A). Review the Theory section of Experiment 9 before attempting this step. Show the logic network in the space provided.

Table 18-3 Decoder Design Problem

Inputs		Outputs			
B	A	Y0	Y1	Y2	Y3
0	0	1	0	0	0
0	1	0	1	0	0
1	0	0	0	1	0
1	1	0	0	0	1

Step 13.   Wire the logic circuit drawn in Step 12 on the computer screen using Electronics Workbench Multisim (or hardwire it) and verify that it matches the truth table in Table 18-3. Use logic probe lights to monitor the outputs and logic switches for inputs A and B.

Question: Did your logic circuit satisfy the requirements of the truth table in Table 18-3? If not, what changes are required?

Name_____

Date_____

19 Multiplexers and Demultiplexers

Objectives:

1. Investigate the operation of an 8-input multiplexer.
2. Investigate the operation of a quad 2-input multiplexer.
3. Demonstrate how a multiplexer can be used to transmit serial data.
4. Demonstrate how a multiplexer can be used as a logic function generator to simulate various logic gates and logic circuits.
5. Demonstrate how a decoder is used as a demultiplexer.
6. Demonstrate an application of a multiplexer and demultiplexer.

Materials:

One 5 V dc power supply
Eleven logic switches
Eleven logic probe lights
One pulse generator
One logic analyzer or oscilloscope
One eight-input multiplexer (1-74151 IC)
One quad two-input multiplexer (1-74157 IC)
One decoder/demultiplexer (1-74138 IC)

Theory:

A **digital multiplexer (data selecter)** is a logic circuit that accepts a number of digital data inputs from several sources and connects one of them to a single output. The multiplexer inputs and output may be **single lines** each or **multiple lines** each. The selection of which digital input will be connected to the output is determined by the binary code applied to the SELECT input. Some multiplexers have one or more **enable inputs** that are used to enable or disable the multiplexer. A multiplexer acts like a **digitally controlled multiposition switch**.

The 74151 in Figure 19-1 is an **8-input multiplexer** with **single line inputs (D7–D0)** and a **single line output (Y)**. It also has an **active low enable input (G)** and an **inverted output (W)**. The active low enable input (G) must be low (connected to ground) to enable the 74151 multiplexer. The SELECT inputs (CBA) require a binary number to select which input will be connected to the output. For example, if binary zero (000) is applied to the SELECT input (CBA), input D0 will connect to the output (Y).

The circuit in Figure 19-2 will demonstrate a 74157 **quad 2-input multiplexer** with **two 4-bit inputs (A and B)** and a **4-bit output (Y).** The two 4-bit inputs are connected to two banks of four switches labeled Code A and Code B, which are applying two 4-bit input codes to the multiplexer. The four output lines are connected to four logic probe lights to monitor the 4-bit multiplexer output. The **SELECT input (A/B)** is only a **single line** because only one of two 4-bit inputs needs to be selected at one time. When a **logic zero (0)** is placed on the SELECT input (A/B), 4-bit **input A** is connected to the 4-bit output. When a **logic one (1)** is placed on the SELECT input (A/B), 4-bit **input B** is connected to the 4-bit output. The 74157 has an **active low enable input (G)** that must be connected to ground (0) to enable the multiplexer.

The circuit in Figure 19-3 will demonstrate how a multiplexer is used as a **parallel-to-serial converter** that transmits a parallel binary word over a single line one bit at a time. The word generator is being used as a **3-bit binary counter** that counts from zero (000) to seven (111) at a frequency of 1 kHz. This count is being applied to the **multiplexer SELET input (CBA)** and will cause a new bit to be output every **1 millisecond**, with **bit 0 being transmitted first** and **bit 7 transmitted last**. Therefore, an entire 8-bit input will be transmitted from the multiplexer output in 8 milliseconds. The binary word transmitted will depend on the settings of switches 0–7. The 74151 multiplexer output is being monitored on the logic analyzer screen.

A multiplexer can be wired as a **logic function generator** to **simulate a combinational logic circuit or a logic gate** directly from the truth table without the need for logic simplification. When wired as a logic function generator, the multiplexer SELECT inputs (CBA) are used as the logic circuit inputs and the multiplexer output (Y) is used as the logic circuit output. The multiplexer data inputs (D0–D7) are connected high (1) or low (0), as necessary to satisfy the logic circuit or logic gate truth table. Therefore the logic circuit or logic gate simulated by the multiplexer can be changed by simply changing the multiplexer data inputs. In order to determine the proper multiplexer data inputs (D0–D7) for simulating a particular logic circuit or logic gate, a truth table must be plotted. Once the truth table has been established, the multiplexer data inputs (D0–D7) that represent those binary inputs (CBA) that produce a binary one (1) output (Y) on the truth table should be set to binary one (1), and those binary inputs (CBA) that produce a binary zero (0) output (Y) on the truth table should be set to binary zero (0). For example, if binary 000 for inputs CBA produces a binary one (1) for output Y on the truth table, data input D0 on the multiplexer should be high (1).

The circuit in Figure 19-4 is a 74151 multiplexer wired as a **3-input logic gate or logic circuit**. The multiplexer SELECT inputs (CBA) represent the **logic gate or circuit inputs (CBA)** and the multiplexer output (Y) represents the **logic gate or circuit output (Y)**. The logic gate or circuit simulated by the multiplexer depends on the settings of switches 0–7, which control the inputs (D0–D7) on the multiplexer.

Demultiplexing is the opposite process from multiplexing. A **demultiplexer** accepts digital data from a **single input** and connects it to one of **many outputs**. The demultiplexer input and outputs may be **single lines** each or **multiple lines** each. The selection of which output will be connected to the input is determined by the binary code applied to the SELECT inputs. Some demultiplexers have an **enable input** that may be used to enable or disable it. Similar to a multiplexer, a demultiplexer acts like a **digitally controlled multiposition switch**.

A **decoder** can be wired as a **demultiplexer** if it has an **enable input**. For this reason, IC chip manufacturers often call these devices **decoder/demultiplexers**. To use a decoder as a demultiplexer, the decoder inputs are used as the demultiplexer SELECT inputs, one of the decoder enable inputs is used as the demultiplexer input, and the decoder outputs are used as the demultiplexer outputs.

The 74138 **decoder/demultiplexer** in Figure 19-5 is wired as a **demultiplexer**. Notice that inputs CBA are being used as the SELECT inputs, the **active low enable input G2A** is being used as the **data input**, and the **active low outputs Y0–Y7** are being used as the **data outputs**. **Active low enable G2B** is connected to ground (0) and **active high enable G1** is connected to +5 V (1). When a binary code is applied to the SELECT input (CBA), and a low (0) is applied to input G2A (active low enable), the 74138 is enabled and the code on the SELECT input (CBA) determines which output goes low (0). When a high (1) is applied to input G2A (active low enable), the 74138 is disabled and the outputs all go high (1), including the output selected by the binary code on the SELECT input (CBA). Therefore, the output selected will follow the input on terminal G2A, causing the decoder to act like a demultiplexer. The remaining outputs will remain high (1).

The circuit in Figure 19-6 is demonstrating a **multiplexer/demultiplexer monitoring system**. Eight switches are being monitored at a remote location and only four lines are connecting the location of the switches to the location of the monitoring panel. The 74151 **multiplexer** is monitoring the eight switches and sends a switch position (one or zero) over a single line to the 74138 **demultiplexer input** (G2A) at the remote monitoring location. The particular switch that is being monitored is determined by the select input (CBA) to the multiplexer and the demultiplexer.

Figure 19-1 Eight-input Multiplexer

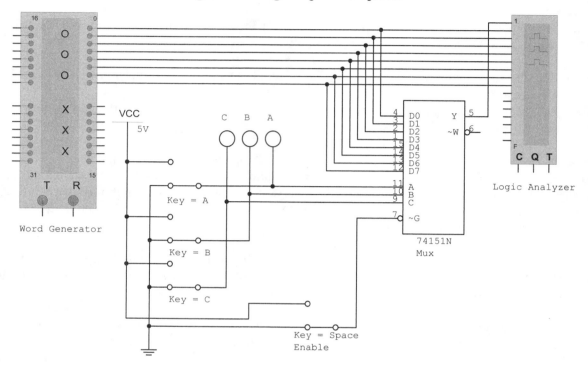

Figure 19-2 Quad 2-input Multiplexer

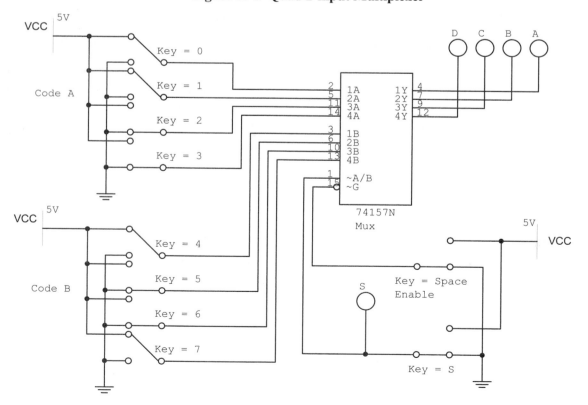

Figure 19-3 Serial Data Transmission

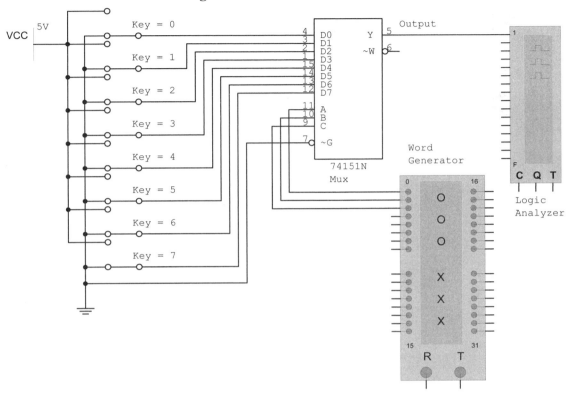

Figure 19-4 Logic Function Generator

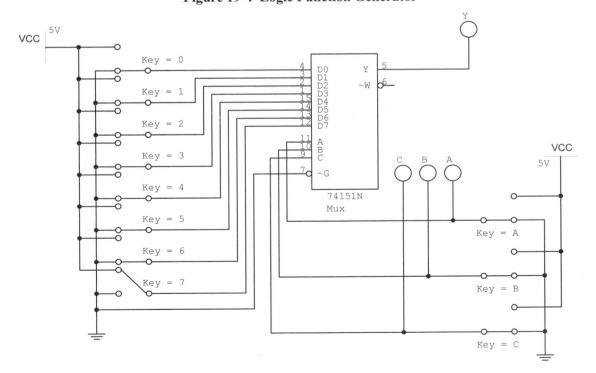

Figure 19-5 One-Line-to-8-Line Demultiplexer

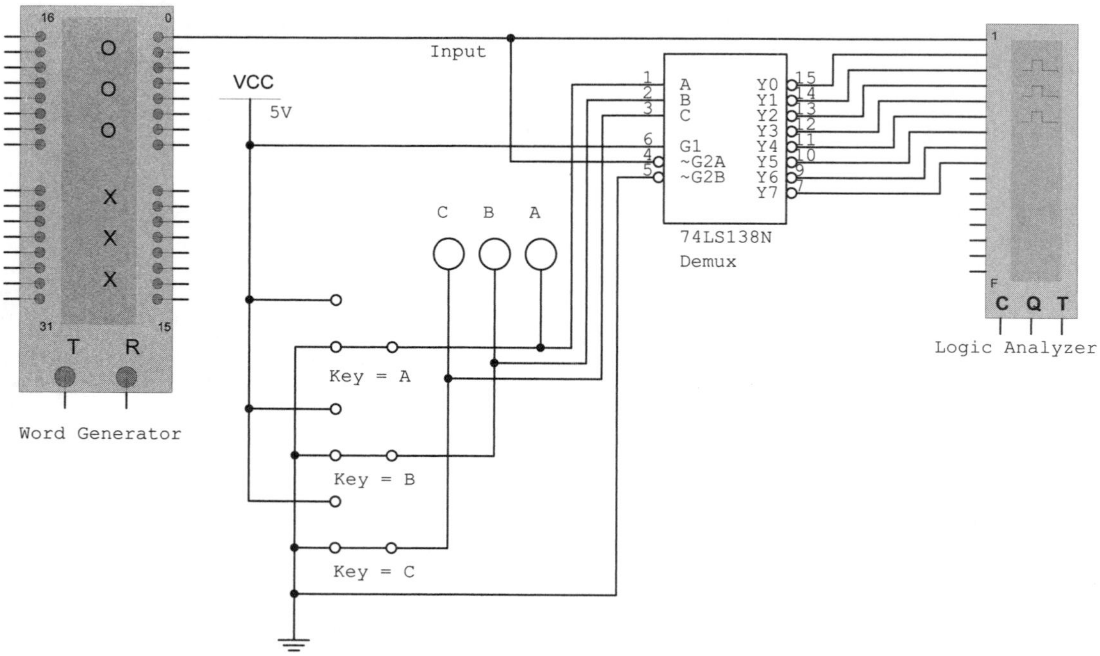

Figure 19-6 Multiplexer/Demultiplexer Monitoring System

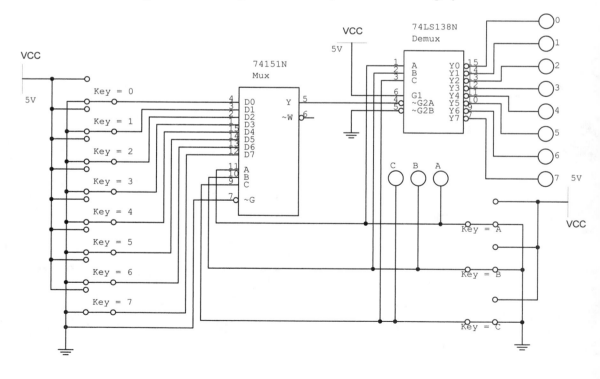

Procedure:

Step 1.   Open circuit file FIG19-1. You are looking at a 74151 8-input multiplexer connected to a word generator, which is applying a different bit pattern to each multiplexer input (D0–D7). The logic analyzer is monitoring the multiplexer output (Y) and inputs D0–D7. Logic switches A, B, and C should be down (0) and the ENABLE switch should be down (G′ = 0). Bring down the word generator enlargement and make sure that the following settings are selected: Frequency = 1 kHz, Trigger = Internal, Controls = Cycle, Hex codes (00000000, 00000001, 00000002, 00000007, 0000000C, 0000001D, 0000003A, 0000007B, 000000F0, 000000F5, 000000E6, 000000E7, 000000C8, 000000C9, 0000008A, 0000008B). Remove the word generator enlargement by clicking the x in the upper right corner. Bring down the logic analyzer enlargement and make sure that the following settings are selected: Clocks/Div = 16, Clock Setup (Clock Source = Internal, Clock Rate = 10 kHz, Pre-trigger Samples = 100, Post-trigger Samples = 1000, Threshold Voltage = 2.5), Trigger Settings (Trigger Clock Edge = Positive, Trigger Qualifier = x, Pattern A= xxxxxxxxxxxxxxxx, Trigger Combinations = A). Move the logic analyzer to the right to make switches A, B, and C and logic probe lights A, B, and C visible by clicking and dragging the top of the analyzer.

Step 2.   Click the On-Off switch to run the simulation. Notice that the waveshape at each multiplexer input (D0–D7) is different. The top waveshape is the multiplexer output, the second waveshape from the top (red) is multiplexer input D0, and the bottom waveshape (blue) is multiplexer input D7. Notice that the output waveshape (top) is the same as the waveshape for input D0 (red) because the multiplexer SELECT input (CBA) is zero (000). By pressing the A, B, and C keys on the computer keyboard, change the SELECT input (CBA) to values between 000 and 111 and determine which input waveshape (D0–D7) is the same as the output waveshape (top) for each SELECT input. Record your answers in Table 19-1. (If the switches do not switch, click the mouse arrow in the circuit window before typing the switch letter on the computer keyboard.) (If the simulation stops, click "Reset" on the logic analyzer to start it again.) **After the table is complete, stop the simulation.**

NOTE: If you are performing this experiment in a hardwired laboratory, use a 1 kHz pulse generator in place of the word generator. Connect it to each multiplexer input to determine which input the output is following for each SELECT input (CBA). Monitor the multiplexer output with an oscilloscope.

Table 19-1 Eight-input Multiplexer

C	B	A	Input (D0–D7)
0	0	0	D0
0	0	1	D1
0	1	0	D2
0	1	1	D3
1	0	0	D4
1	0	1	D5
1	1	0	D6
1	1	1	D7

Question: What conclusion can you draw about the relationship between the multiplexer output and the inputs based on the SELECT inputs (CBA)?

THEY'RE BINARY SAME NUMBER TO DECIMAL

Step 3.   Start the simulation again, click the mouse in the circuit window, and set the ENABLE switch to binary one (up) by pressing the space bar on the computer keyboard. This places a binary one (1) on the 74151 multiplexer active low enable terminal (G' = 1). Stop the simulation after one complete screen display.

Question: What do you notice about the 74151 multiplexer output waveshape (top waveshape)? **Explain your answer.**

IT'S FLAT

Step 4.   Open circuit file FIG19-2. You are looking at a 74157 quad 2-input multiplexer wired to
            two banks of four switches, which are applying two 4-bit binary codes to the two 4-bit
            multiplexer inputs. Notice that one 4-bit code (Code A) is equal to binary three (0011) and
            the other 4-bit code (Code B) is equal to binary nine (1001). Four logic probe lights are
            monitoring the 4-bit multiplexer output. Select switch S should be down (A/B = 0) and the
            ENABLE switch should be down (G' = 0). Click the On-Off switch to run the simulation
            and observe the output probe lights, then stop the simulation.

Question: Is the 4-bit binary code connected to the multiplexer A input (Code A) or the 4-bit binary
code connected to the B input (Code B) passing through to the multiplexer output?

Code A

Step 5.   Run the simulation and press the S key on the keyboard to set the select input to binary one
            (1). Observe the multiplexer outputs, then stop the simulation.

Questions: Is the 4-bit binary code connected to the multiplexer A input (Code A) or the 4-bit binary
code connected to the B input (Code B) passing through to the multiplexer output?

Code B

What conclusion can you draw about the relationship between the multiplexer input code and the output
code for each select (S) setting?

if S=1 ⇒ A enabled
   S=0 ⇒ B enabled

Step 6.   Run the simulation and press the SPACE bar on the keyboard to set the ENABLE switch to
            binary one (G' = 1) and observe the multiplexer outputs, then stop the simulation.

Questions: What happened to the multiplexer outputs?

All 1

What conclusion can you draw about the 74157 multiplexer when the active low ENABLE (G') input is
high (1)?

Step 7.   Open circuit file FIG19-3. You are looking at a 74151 multiplexer wired as a parallel-to-serial converter that transmits a parallel binary code over a single output line one bit at a time. The word generator is being used as a 3-bit binary counter that counts from zero (000) to seven (111) at a frequency of 1 kHz. This count is being applied to the multiplexer SELECT input (CBA) and will cause a new bit to be output every 1 millisecond. Therefore, an entire 8-bit input will be transmitted from the multiplexer output in 8 milliseconds. The binary code transmitted will depend on the settings of switches 0–7. Bring down the word generator enlargement and make sure that the following settings are selected: Frequency = 1 kHz, Trigger = Internal, Controls = Burst, Hex codes (00000000, 00000001, 00000002, 00000003, 00000004, 00000005, 00000006, 00000007). Remove the word generator enlargement by clicking the x in the upper right corner of the generator. Bring down the logic analyzer enlargement and make sure that the following settings are selected: Clocks/Div = 16, Clock Setup (Clock Source = Internal, **Clock Rate = 16 kHz,** Pre-trigger Samples = 100, Post-trigger Samples = 1000, Threshold Voltage = 2.5), Trigger Settings (Trigger Clock Edge = Positive, Trigger Qualifier = x, Pattern A = xxxxxxxxxxxxxxxx, Trigger Combinations = A). Move the logic analyzer to the right so that the logic switches 0–7 are visible. These switches can be changed by pressing the equivalent number key on the computer keyboard after clicking the mouse arrow in the circuit window. Set the switches (0–7) to any binary code other than zero. Click the On-Off switch to run the simulation. After the display is complete, stop the simulation. The multiplexer output is being monitored on the logic analyzer screen, with bit 0 being transmitted first and bit 7 transmitted last. Each division on the logic analyzer screen is one bit time.

NOTE: If this part of the experiment is being performed in a hardwired laboratory, replace the word generator with a 3-bit binary counter clocked at a frequency of 1 kHz. Replace the logic analyzer with an oscilloscope.

Question: Is the transmitted waveshape on the logic analyzer screen as expected for the switch settings?

NOTE: If you wish to try another code **when using Multisim 7**, you must first reset the cursor to the top of the number column on the word generator by right clicking the top number and selecting "Set Cursor".

Step 8.   Open circuit file FIG19-4. You are looking at a 74151 8-input multiplexer wired as a 3-input logic function generator simulating a 3-input logic gate or circuit with inputs CBA and output Y. The multiplexer SELECT inputs (CBA) represent the logic gate or circuit inputs (CBA) and the multiplexer output (Y) represents the logic gate or circuit output (Y). The logic gate or circuit simulated by the multiplexer depends on the setting of switches 0–7 (binary inputs to D0–D7 on the multiplexer). Switch 7 should be up (1) and switches 0–6 should be down (0). Click the On-Off switch to run the simulation. By pressing the A, B, and C keys on the keyboard, change logic inputs A, B, and C to the values in Table 19-2 and record the output (Y) for each input combination. **After the table is complete, stop the simulation.**

Table 19-2 Truth Table

C	B	A	Y
0	0	0	
0	0	1	
0	1	0	
0	1	1	
1	0	0	
1	0	1	
1	1	0	
1	1	1	

Question: Based on the results in Table 19-2, what 3-input logic gate is being simulated by the multiplexer circuit?

Step 9.   Plot the truth table for a 3-input OR gate in Table 19-3.

Table 19-3 Truth Table

C	B	A	Y
0	0	0	
0	0	1	
0	1	0	
0	1	1	
1	0	0	
1	0	1	
1	1	0	
1	1	1	

Step 10.   Change switches 0–7 to simulate the OR gate truth table plotted in Table 19-3. Click the On-Off switch to run the simulation. By pressing the A, B, and C keys on the keyboard, test your circuit for all possible input combinations to see if your circuit matches the truth table (Table 19-3), then stop the simulation.

Question: Did your results indicate that you simulated a 3-input OR gate? If not, what changes are needed?

Step 11.   Plot the truth table for a 3-input NAND gate in Table 19-4.

Table 19-4 Truth Table

C	B	A	Y
0	0	0	
0	0	1	
0	1	0	
0	1	1	
1	0	0	
1	0	1	
1	1	0	
1	1	1	

Step 12.   Change switches 0–7 to simulate the NAND gate truth table plotted in Table 19-4. Click
            the On-Off switch to run the simulation. By pressing the A, B, and C keys on the keyboard,
            test your circuit for all possible input combinations to see if your circuit matches the truth
            table (Table 19-4), then stop the simulation.

Question: Did your results indicate that you simulated a 3-input NAND gate? If not, what changes are
needed?

Step 13.   If you have not completed Experiment 6, complete it now before continuing this experiment.

Step 14.   Change switches 0–7 to simulate the logic circuit truth table plotted for the circuit in Figure
            6-4, Experiment 6. Click the On-Off switch to run the simulation. By pressing the A, B, and
            C keys on the keyboard, test your circuit for all possible input combinations to see if your
            circuit matches the truth table for the logic circuit in Figure 6-4, then stop the simulation.

Question: Did your results indicate that you simulated the logic circuit in Figure 6-4, Experiment 6?
If not, what changes are needed?

Step 15.   Open circuit file FIG 19-5. You are looking at a 74138 3-line-to-8-line decoder wired as a 1-line-to-8-line demultiplexer. Notice that inputs CBA are being used as the SELECT inputs, the active low enable input G2A is being used as the data input, and the active low outputs Y0–Y7 are being used as the data outputs. The logic analyzer is monitoring the input (G2A) and outputs Y0–Y7. Logic switches A, B, and C should be down (0). Bring down the word generator enlargement and make sure that the following settings are selected: Frequency = 1 kHz, Trigger = Internal, Controls = Cycle, Hex codes (00000000, 00000001, repeated 8 times). Remove the word generator enlargement by clicking the x in the upper right corner. Bring down the logic analyzer enlargement and make sure that the following settings are selected: Clocks/Div = 16, Clock Setup (Clock Source = Internal, Clock Rate = 10 kHz, Pre-trigger Samples = 100, **Post-trigger Samples = 100000**, Threshold Voltage = 2.5), Trigger Settings (Trigger Clock Edge = Positive, Trigger Qualifier = x, Pattern A= xxxxxxxxxxxxxxxx, Trigger Combinations = A). Move the logic analyzer to the right to make switches A, B, and C and logic probe lights A, B, and C visible by clicking and dragging the top of the analyzer. See the Theory section for further details.

Step 16.   Click the On-Off switch to run the simulation. The top waveshape on the logic analyzer screen is the demultiplexer input, the second waveshape from the top (red) is output Y0, and the bottom waveshape (blue) is output Y7. Notice that the output Y0 waveshape (red) is identical to the input waveshape (top) because the SELECT input (CBA) is binary zero (000). By pressing the A, B, and C keys on the keyboard, change the select input (CBA) to values between zero (000) and seven (111) and determine which output waveshape is identical to the input waveshape (top). Record your results in Table 19-5. Click the arrow in the circuit window before pressing the A, B, or C keys. (If the simulation stops, click "Reset" on the logic analyzer to start it again.) **After the table is complete, stop the simulation.**

Table 19-5 1-Line-to-8-Line Multiplexer

C	B	A	Output (Y0–Y7)
0	0	0	
0	0	1	
0	1	0	
0	1	1	
1	0	0	
1	0	1	
1	1	0	
1	1	1	

NOTE: If this part of the experiment is performed in a hardwired laboratory, use a 1 kHz pulse generator in place of the word generator. If a logic analyzer is not available, monitor each demultiplexer output (Y0–Y7) with an oscilloscope to determine which output is active for different SELECT inputs (CBA).

Question: What conclusion can you draw about the relationship between the demultiplexer outputs and the input, based on the SELECT inputs (CBA)?

Step 17.   Open circuit file FIG19-6. You are looking at a multiplexer/demultiplexer monitoring system that will demonstrate an application for a multiplexer and a demultiplexer. Eight switches are being monitored at a remote location without running eight separate lines between the switches and the remote monitoring panel. For a more detailed discussion of the system, see the Theory section.

Step 18.   Click the On-Off switch to run the simulation. Press the number keys on the computer keyboard to switch the numbered switches on and off (up and down) to determine which switch is being monitored by the logic probe lights.

Question: Which switch is being monitored by the logic probe lights? **Explain why that particular switch is being monitored.**

Step 19.   Change the SELECT input (CBA) to another binary code by pressing the A, B, and C keys on the keyboard. Press the number keys again to switch the numbered switches on and off (up and down) to determine which switch is being monitored by the logic probe lights, then stop the simulation.

Question: Is the correct switch being monitored by the logic probe lights?

EXPERIMENT

20 Troubleshooting MSI Logic Circuits

Objectives:

1. Determine the defective component in an encoder/decoder decimal display system.
2. Determine the defective component in a multiplexer/decoder decimal display system.
3. Determine the defective component in a multiplexer/demultiplexer security monitoring system.

Materials:

This experiment can only be performed on Electronics Workbench Multisim using the circuits disk provided with this manual.

Theory:

In order to perform this experiment effectively, you must first complete Experiments 18–19. Use the theory learned in those experiments to find the defective components in the logic circuits in this experiment.

Before attempting to find the defective component, test the circuit for all possible input conditions to determine which input conditions are producing the incorrect outputs. This should give you some direction in determining where to begin testing. While the input is producing an incorrect output, use the logic probe light to measure logic levels at various test points to help determine the defective component.

Procedure:

1. Open circuit file FIG20-1. Click the On-Off switch to run the simulation. This circuit should display the decimal number of the highest numbered switch that is down (0) on the LED display. Using the logic probe light to measure the logic levels at various test points in the circuit, determine which component is defective. To switch the numbered switches, press the key on the computer keyboard that matches the number label on the switch.

    Defective component _____

2.   Open circuit file FIG20-2. Click the On-Off switch to run the simulation. This circuit should display the decimal number of the highest numbered switch that is down (0) on the LED display. Using the logic probe light to measure the logic levels at various test points in the circuit, determine which component is defective. To switch the numbered switches, press the key on the computer keyboard that matches the number label on the switch.

        Defective component _____

3.   Open circuit file FIG20-3. Click the On-Off switch to run the simulation. This circuit should display the decimal number of the highest numbered switch that is down (0) on the LED display. Using the logic probe light to measure the logic levels at various test points in the circuit, determine which component is defective. To switch the numbered switches, press the key on the computer keyboard that matches the number label on the switch.

        Defective component _____

4.   Open circuit file FIG20-4. Click the On-Off switch to run the simulation. This circuit should display the decimal number for input Code A when S = 0 or the decimal number for input Code B when S = 1. Using the logic probe light to measure the logic levels at various test points in the circuit, determine which component is defective.

        Defective component _____

5.   Open circuit file FIG20-5. Click the On-Off switch to run the simulation. This circuit should display the decimal number for input Code A when S = 0 or the decimal number for input Code B when S = 1. Using the logic probe light to measure the logic levels at various test points in the circuit, determine which component is defective.

        Defective component _____

6.   Open circuit file FIG20-6. Click the On-Off switch to run the simulation. The code on the SELECT input (CBA) should determine which switch is being monitored. The remaining logic lights should be "on" (1) continuously. To switch the switches, press the key on the computer keyboard that matches the number or letter label on the switch. Using the logic probe light to measure the logic levels at various test points in the circuit, determine which component is defective.

        Defective component _____

7.   Open circuit file FIG20-7. Click the On-Off switch to run the simulation. The code on the SELECT input (CBA) should determine which switch is being monitored. The remaining logic lights should be "on" (1) continuously. To switch the switches, press the key on the computer keyboard that matches the number or letter label on the switch. Using the logic probe light to measure the logic levels at various test points in the circuit, determine which component is defective.

        Defective component _____

8.   Open circuit file FIG20-8. Click the On-Off switch to run the simulation. The code on the
        SELECT input (CBA) should determine which switch is being monitored. The remaining logic
        lights should be "on" (1) continuously. To switch the switches, press the key on the computer
        keyboard that matches the number or letter label on the switch. Using the logic probe light to
        measure the logic levels at various test points in the circuit, determine which component is
        defective.

        Defective component _____

Sequential Logic Circuits

The logic circuits studied thus far have been **combinational logic circuits**, which have output responses that follow changes in the input levels with **minimum time delay**. Prior input conditions have no effect on the present output levels because combinational logic circuits do not have memory.

The experiments in Part IV involve the study of **sequential logic circuits**. Sequential logic circuits have present outputs that are affected by prior input levels because they have **memory**. The basic building blocks of sequential logic circuits are **flip-flops** and **latching circuits**. In the first two experiments, you will study how latches and flip-flops store binary data. Next, you will study how **monostable and astable multivibrators (pulse oscillators)** generate pulses. In the remaining experiments, you will study **registers** and **counters** and learn how they are wired using basic flip-flops. In the final experiment you will **troubleshoot** some sequential logic circuits involving latches, flip-flops, registers, and counters.

The circuits for the experiments in Part IV can be found on the enclosed disk in the PART4 subdirectory.

EXPERIMENT

21

S-R and D Latches

Objectives:

1.   Investigate the operation of a NOR gate S-R latch.
2.   Investigate the operation of a NAND gate S-R latch.
3.   Investigate the operation of a D latch.

Materials:

One 5 V dc power supply
Two logic switches
Four logic probe lights
Two two-input NOR gates (1-7402 IC)
Four two-input NAND gates (1-7400 IC)
One INVERTER (1-7404 IC)
One D-latch (1-7475 IC)
Two square wave generators
One logic analyzer or dual-trace oscilloscope

Theory:

A **latch** is a **bistable storage device** with an output that can be latched into **one of two states**. This is caused by a feedback arrangement in which the outputs of two logic gates are fed back to the opposite gate inputs.

The most basic **binary data storage device** is the **set-reset (S-R) latch**. The S-R latch has two inputs, **set (S)** and **reset (R)**. The **output (Q)** will latch high (1) when the set (S) input is active and will latch low (0) when the reset (R) input is active.

A **D latch** has only **one input**. The D latch also has an **enable (EN)** terminal, which is used to control whether the latch accepts or ignores the input (D). When the latch is enabled, the output (Q) will follow the input (D). When the latch is disabled, the output (Q) will **"latch" (store)** the last input level (D) before it was disabled. Because the output (Q) follows the input (D) when the latch is enabled, a D latch is often referred to as a **"transparent" latch**.

The logic circuit in Figure 21-1 is a set-reset (S-R) latch wired with two NOR gates and is called a **NOR gate S-R latch**. When the set (S) input is high (1) and the reset (R) input is low (0), the output (Q) will become latched in the high state (Q = 1). When both the set (S) and reset (R) inputs are low (0), the output (Q) stays latched at its previous state. When the reset (R) input is high (1) and the set (S) input is low (0), the output (Q) will become latched in the low state (Q = 0). The Q′ output will always be the inverse of the Q output, except when both the **set (S) and reset (R) inputs** are **high (1)**, which is a **"not allowed"** input condition for a NOR gate latch.

The logic circuit in Figure 21-2 is a set-reset (S-R) latch wired with two NAND gates and is called a **NAND gate S-R latch**. When the set (S′) input is low (0) and the reset (R′) input is high (1), the output (Q) will become latched in the high state (Q = 1). When both the set (S′) and reset (R′) inputs are high (1), the output (Q) stays latched at its previous state. When the reset (R′) input is low (0) and the set (S′) input is high (1), the output (Q) will become latched in the low state (Q = 0). Therefore, the set (S′) and reset (R′) inputs of a NAND latch are **active low inputs**. The Q′ output will always be the inverse of the Q output, except when both the **set (S′) and reset (R′) inputs** are **low (0)**, which is a **"not allowed"** input condition for a NAND latch gate.

The logic circuit in Figure 21-3 is a **D latch** wired with four 2-input NAND gates and an INVERTER. When the **enable input (EN) is low (0)**, latch inputs S′ and R′ are both high (1). This causes the S-R latch to stay "latched" on the previous input and not respond to any D inputs. When the **enable input (EN) is high (1)**, the latch inputs S′ and R′ depend on whether input D is high (1) or low (0). If **input D is high (1)**, latch input S′ is low (0) and input R′ is high (1), causing the latch to set and the output to go high (Q = 1). If **input D is low (0)**, latch input S′ is high (1) and input R′ is low (0), causing the latch to reset and the output to go low (Q = 0). Therefore, the latch output (Q) will follow the input (D) when the enable (EN) terminal is high (1) and the latch output (Q) will stay "latched" when the enable (EN) terminal is low (0).

The 7475 in Figure 21-4 has **four D latches** that are similar to the D latch in Figure 21-3. The enable terminals for latches 1 and 2 are connected together (EN1) and the enable terminals for latches 3 and 4 are connected together (EN2). In this experiment only latches 1 and 2 are shown, and you will use only latch 1. When the enable terminal (EN1) is high (1), the output (Q) will follow the input (D1) and the latch is **"transparent."** When the enable terminal (EN1) is low (0), the output (Q) is **"latched."**

The circuit in Figure 21-5 will monitor the **timing** of the **7475 D latch**. The word generator will apply a pulse waveshape to the input (D1) and a different pulse waveshape to the enable input (EN1). The logic analyzer will monitor the input (D1), the enable input (EN1), and the output (Q1). When the enable input (EN1) is high (1), the D latch is "transparent" and the output (Q1) will follow the input (D1). When the enable input (EN1) is low (0), the D latch output is "latched" on the value of the input (D1) when the enable went low (0).

Figure 21-1 NOR Gate S-R Latch

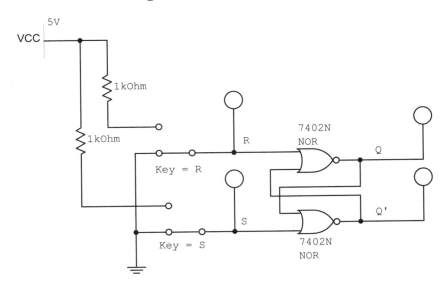

Figure 21-2 NAND Gate S-R Latch

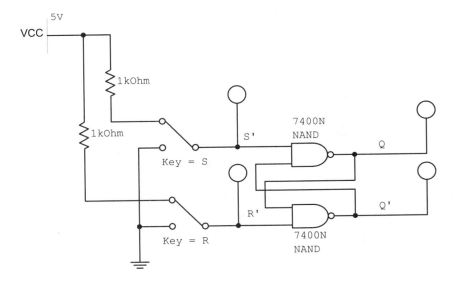

Figure 21-3 NAND Gate D Latch

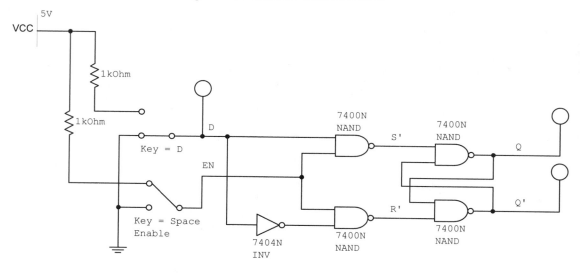

Figure 21-4 7475 D Latch

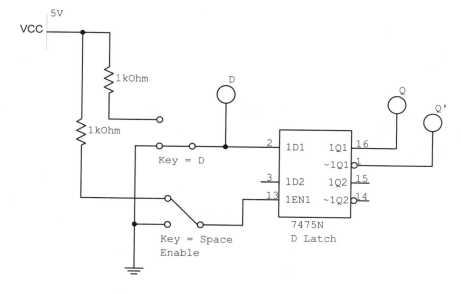

Figure 21-5 7475 D Latch Timing

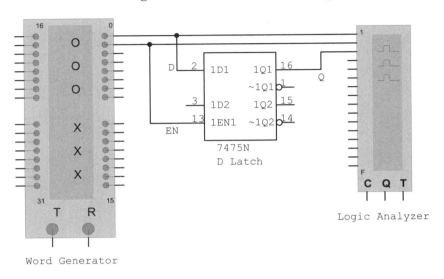

Procedure:

Step 1.   Open circuit file FIG 21-1. You are looking at an S-R latch wired with two 2-input NOR gates. Both the S and R switches should be down (0). Click the On-Off switch to run the simulation. Press the S key on the computer keyboard to bring the set (S) input to binary one (1).

NOTE: When this experiment is performed in a hardwired laboratory, the output (Q) will set (1) or reset (0) when power is turned on. This does not happen in the computer simulation because there is no power-up transient to trigger the latch.

Question: What are outputs Q and Q′ when S = 1 and R = 0?

$Q = 1$

$Q' = 0$

Step 2.   Press the S key again to bring the set (S) input to binary zero (0).

Question: What happened to outputs Q and Q′ when S = 0 and R = 0?

$Q = 1$

$Q' = 0$

Step 3.   Press the R key to bring the reset (R) input to binary one (1).

Question: What are outputs Q and Q' when S = 0 and R =1?

$$Q = 0$$
$$Q' = 1$$

Step 4.   Press the R key again to bring the reset (R) input to binary zero (0).

Question: What happened to outputs Q and Q' when S = 0 and R = 0?

$$Q = 0$$
$$Q' = 1$$

Step 5.   Press the S key and the R key to bring both the set (S) and reset (R) inputs to binary one (1) and observe the outputs, then stop the simulation.

Question: What happened to outputs Q and Q' when S = 1 and R = 1? **Explain.**

$$Q = 0$$
$$Q' = 0$$

Step 6.   Open circuit file FIG21-2. You are looking at an S-R latch wired with two 2-input NAND gates. Both the S and the R switches should be up (1). Click the On-Off switch to run the simulation. Press the S key on the computer keyboard to bring the set (S') input to binary zero (0).

NOTE: When this experiment is performed in a hardwired laboratory, the output (Q) will set (1) or reset (0) when power is turned on. This does not happen in the computer simulation because there is no power-up transient to trigger the latch.

Question: What are outputs Q and Q' when S' = 0 and R' = 1?

$$Q = 1$$
$$Q' = 0$$

Step 7.   Press the S key again to bring the set (S') input to binary one (1).

Question: What happened to outputs Q and Q' when S' = 1 and R' = 1?

$Q = 1$

$Q' = 0$

Step 8.   Press the R key to bring the reset (R') input to binary zero (0).

Question: What are outputs Q and Q' when S' = 1 and R' = 0?

$Q = 0$

$Q' = 1$

Step 9.   Press the R key again to bring the reset (R') input to binary one (1).

Question: What happened to outputs Q and Q' when S' = 1 and R' = 1.

$Q = 0$

$Q' = 1$

Step 10.   Press the S key and the R key to bring both the set (S') and reset (R') inputs to binary zero (0) and observe the outputs, then stop the simulation.

Question: What happened to outputs Q and Q' when S' = 0 and R' = 0? **Explain.**

$Q = 1$

$Q' = 1$ } IT'S A DOUBLE JAM. S AND R SHOULDN'T BE 1 AT THE SAME TIME

Step 11.   Open circuit file FIG 21-3. You are looking at a D latch wired with four 2-input NAND gates and an INVERTER. See the Theory section for a detailed discussion of this logic network. Switch D should be down (0) and the ENABLE switch should be up (1). Click the On-Off switch to run the simulation.

Question: What are outputs Q and Q' with D = 0?

$Q = 0$

$Q' = 1$

Step 12.   Press the D key to set the D input to binary one (1).

Question: What happened to outputs Q and Q' with D = 1?

$$Q = 1$$
$$Q' = 0$$

Step 13.   Press the D key again to clear the D input to binary zero (0).

Question: What happened to output Q? **Explain.**

Q = 0 AGAIN

Q Follows D, when enabled

Step 14.   Press the space bar to bring the enable (EN) terminal to binary zero (0). Press the D key to set the D input to binary one (1) and observe the outputs, then stop the simulation.

Question: What happened to output Q? **Explain.**

Step 15.   Open circuit file FIG21-4. You will test a 7475 D latch and compare the results with the results for the NAND gate D latch in Figure 21-3. Switch D should be down (0) and the ENABLE switch should be up (1). Click the On-Off switch to run the simulation.

Questions: What is output Q with D = 0?

$$Q = 0$$

How do your results compare with the results for the NAND gate D latch in Figure 21-3?

SAME

Step 16.   Press the D key to set the D input to binary one (1).

Questions: What is output Q with D = 1?

$$Q = 1$$

How do your results compare with the results for the NAND gate D latch in Figure 21-3?

SAME

Step 17.   Press the space bar to bring the enable terminal (EN1) to binary zero (0). Press the D key to clear the D input to binary zero (0) and observe the outputs, then stop the simulation.

Questions: What happened to output Q? **Explain.**

IT DON'T CHANGE. BECAUSE it WAS DISABLED, SO NO CHANGES.

How do your results compare with the results for the NAND gate D latch in Figure 21-3?

SAME RESULTS

Step 18.   Open circuit file FIG21-5. You are looking at a circuit for measuring the timing of the 7475 D latch. Bring down the word generator enlargement and make sure that the following settings are selected: Frequency = 1 kHz, Trigger = Internal, Controls = Burst, Hex codes (00000002, 00000003, 00000002, 00000003, 00000000, 00000001, 00000000, 00000001, repeated twice). Bring down the logic analyzer enlargement and make sure that the following settings are selected: Clocks/Div = 16, Clock Setup (Clock Source = Internal, Clock Rate = 10 kHz, Pre-trigger Samples = 100, Post-trigger Samples = 1000, Threshold Voltage = 2.5), Trigger Settings (Trigger Clock Edge = Positive, Trigger Qualifier = x, Pattern A= xxxxxxxxxxxxxxxx, Trigger Combinations = A). Click the On-Off switch to run the simulation. The word generator is applying a pulse waveshape to the D input and a different pulse waveshape to the enable input (EN) of the D latch. The logic analyzer is monitoring input D (red), enable input EN (green), and output Q (blue).

NOTE: If this experiment is performed in a hardwired laboratory, use two pulse generators in place of the word generator and a dual-trace oscilloscope in place of the logic analyzer if a logic analyzer is not available.

Questions: What do you notice about output Q (blue) when enable EN (green) is high (1)?

IT FOLLOWS D

What do you notice about output Q (blue) when enable EN (green) is low (0)?

IT STAYS AS iT WAS BEFORE. DOES NOT CHANGE.

EXPERIMENT

Edge-Triggered Flip-Flops

Objectives:

1. Investigate the operation of an edge-triggered S-R flip-flop.
2. Investigate the operation of an edge-triggered D flip-flop.
3. Investigate the operation of an edge-triggered J-K flip-flop.

Materials:

One 5 V dc power supply
Five logic switches
Four logic probe lights
One INVERTER (1-7404 IC)
One positive-edge-triggered D flip-flop (1-7474 IC)
Four two-input NAND gates (1-7400 IC)
One negative-edge-triggered J-K flip-flop (1-74112 IC)
Three square wave generators
One logic analyzer or dual-trace oscilloscope
One 0.1 μF capacitor
One 10 kΩ resistor

Theory:

Logic circuits can be either **asynchronous** or **synchronous**. In an asynchronous logic circuit, the output changes state any time an input changes state. In a synchronous logic circuit, the exact time at which the output can change state is determined by a series of rectangular pulses called **clock pulses**. A **flip-flop** is a **synchronous bistable device** because the output will go to a binary one (1) or zero (0) only when the clock input receives a clock pulse. An **edge-triggered flip-flop** output changes state only on the **positive edge** (rising edge) or only on the **negative edge** (falling edge) of the clock pulse, depending on whether it is a **positive-edge-triggered** or a **negative-edge-triggered** flip-flop.

The logic circuit in Figure 22-1 is an S-R latch wired as a **positive-edge-triggered S-R flip-flop**. The capacitor (C) and the resistor (R) form an **edge detector** circuit that provides a short duration pulse at the edge detector output (EN) on the **positive edge** of the clock pulse at the edge detector input (CLK). The edge detector circuit is able to produce the short duration pulse at the output (EN) because of the short charge time of the capacitor. This produces a short duration one (1) output at terminal EN. This will open the pulse steering circuit IC2A and IC2B for a short duration allowing the S-R latch to set (Q = 1), clear (Q = 0), or stay the same based on the inputs at S and R during the clock transition period. When inputs S = 1 and R = 0, the flip-flop output will set (Q = 1) on the positive edge of the clock (CLK) pulse. When inputs S = 0 and R = 1, the flip-flop output will clear (Q = 0) on the positive edge of the clock pulse.

When inputs S = 0 and R = 0, the flip-flop output will not change state on the positive edge of the clock pulse. Inputs S = 1 and R = 1 are "not allowed" input conditions for an S-R flip-flop. The edge detector circuit can be changed to produce a short duration pulse on the negative edge of the clock pulse by placing an INVERTER at the clock input (CLK). This would produce a **negative-edge-triggered flip-flop**.

The logic circuit in Figure 22-2 is an **S-R latch** wired as a **positive-edge-triggered D flip-flop**. It is identical to the positive-edge-triggered S-R flip-flop in Figure 22-1, except an INVERTER (IC1A) is connected between the S and R inputs of the S-R flip-flop to make the D flip-flop. If D = 1, S will be one (1) and R will be zero (0) causing the flip-flop output to set (Q = 1) when the **clock input (CLK)** receives a **positive clock edge**. If D = 0, S will be zero (0) and R will be one (1) causing the flip-flop output to clear (Q = 0) when the clock input (CLK) receives a positive clock edge. The positive-edge-triggered D flip-flop will not respond to changes in the D input until the clock input (CLK) receives a positive clock edge. A **negative-edge-triggered D flip-flop** operates the same as a positive-edge-triggered D flip-flop except it responds to a negative clock edge instead of a positive clock edge. The D flip-flop is often used to **store a single data bit**.

The 7474 in Figure 22-3 has two **positive-edge-triggered D flip-flops** with **active low asynchronous preset (PR) and clear (CLR) inputs**. You will use only one of the D flip-flops in this experiment. The asynchronous inputs override the **CLK (synchronous) input** and will set (Q = 1) or clear (Q = 0) the flip-flop output whenever a binary zero (0) is applied to the PR or CLR active low terminals. These asynchronous inputs are not edge-triggered and respond to dc levels. The S and C switches control the voltage levels (high or low) applied to the asynchronous PR or CLR inputs. Setting both asynchronous PR and CLR terminals low (0) at the same time is a "not allowed" input condition. The output of the flip-flop will set (Q = 1) on the positive edge of the clock pulse when D = 1, and clear (Q = 0) on the positive edge of the clock pulse when D = 0. The flip-flop inverted output (Q') will be the inverse of the output (Q).

The circuit in Figure 22-4 will display the **timing** of the **7474 positive-edge-triggered D flip-flop**. The word generator is applying a series of clock pulses to the clock input (CLK) and a pulse waveshape to the D input of the flip-flop. The logic analyzer is monitoring the CLK input, the D input, the output (Q), and the inverted output (Q').

The 74112 in Figure 22-5 has two **negative-edge-triggered J-K flip-flops** with active low **asynchronous preset (PR) and clear (CLR) inputs**. (The bubble on the CLK input designates that it is negative edge triggered.) You will use only one of the J-K flip-flops in this experiment. The asynchronous inputs on the 74112 operate the same as the asynchronous inputs on the 7474 D flip-flop. (See the discussion of the 7474 D flip-flop.) The S and C switches control the voltage levels (high or low) applied to the asynchronous PR and CLR inputs. The J-K flip-flop is similar to an S-R flip-flop, except the J input is the set input and the K input is the reset (clear) input in a J-K flip-flop. When inputs J = 1 and K = 0, the J-K flip-flop output will set (Q = 1) on the negative clock edge. When inputs J = 0 and K = 1, the J-K flip-flop output will clear (Q = 0) on the negative clock edge. When inputs J = 0 and K = 0, the J-K flip-flop output (Q) will not change state on the negative clock edge. The most important feature of the J-K flip-flop is an additional state with inputs J = 1 and K = 1, which is not allowed in the S-R flip-flop. When inputs J = 1 and K = 1, the J-K flip-flop output will toggle between set (Q = 1) and clear (Q = 0) on each negative clock edge. The J-K flip-flop is the most versatile and widely used type of flip-flop.

The circuit in Figure 22-6 will display the **timing** of the **74112 negative-edge-triggered J-K flip-flop**. The word generator is applying a series of clock pulses to the clock input (CLK), a pulse waveshape to the J input, and a different pulse waveshape to the K input of the flip-flop. The logic analyzer is monitoring the CLK input, the J input, the K input, the output (Q) and the inverted output (Q').

Figure 22-1 Edge-Triggered S-R Flip-Flop

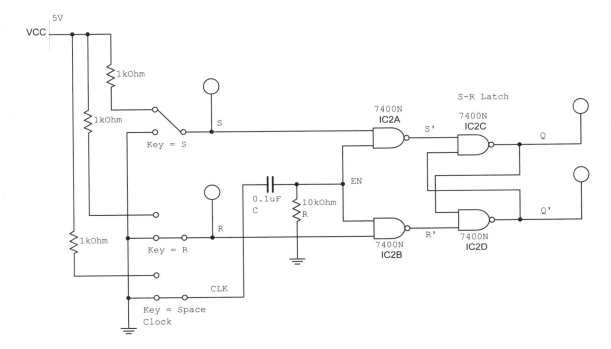

Figure 22-2 Edge-Triggered D Flip-Flop

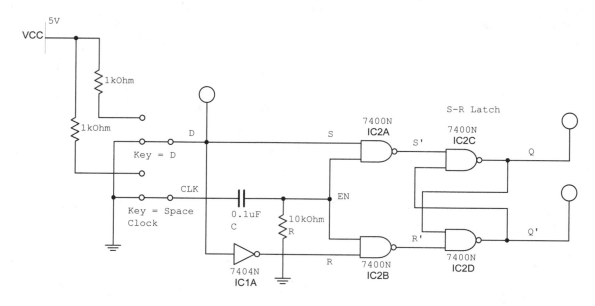

Figure 22-3 7474 Edge-Triggered D Flip-Flop

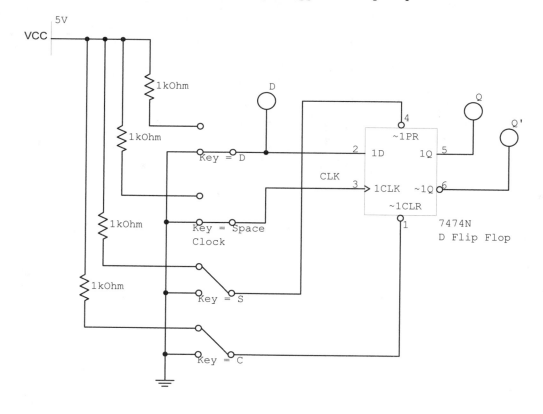

Figure 22-4 7474 Edge-Triggered D Flip-Flop Timing

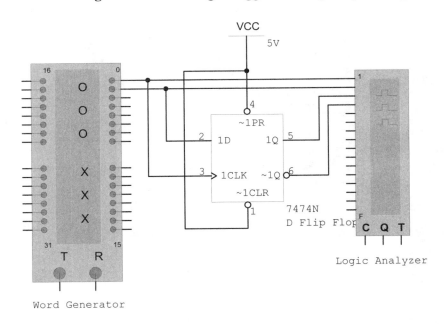

Figure 22-5 74112 Edge-Triggered J-K Flip-Flop

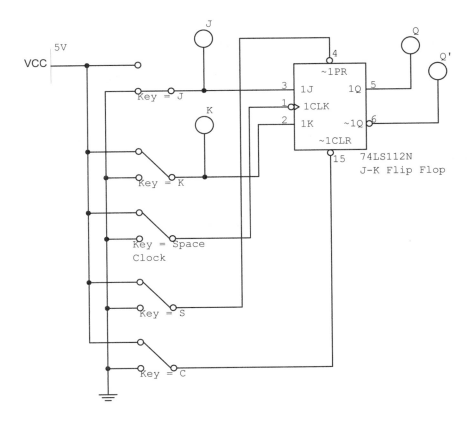

Figure 22-6 74112 Edge-Triggered J-K Flip-Flop Timing

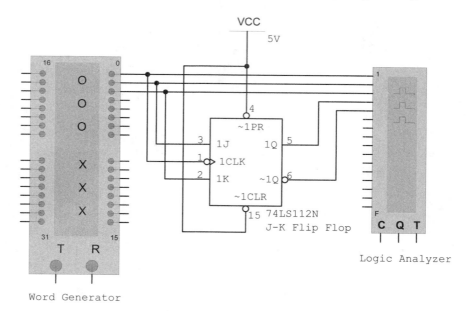

Procedure:

Step 1.   Open circuit file FIG22-1. You are looking at an S-R latch wired as a positive-edge-triggered S-R flip-flop. The capacitor (C) and resistor (R) form an edge detector circuit that provides a short duration pulse at the edge detector output (EN) on the positive edge of the clock pulse at the edge detector input (CLK). See the Theory section for more details about the operation of this circuit.

Step 2.   The S switch should be up (1) and the R and CLOCK switches should be down (0). Click the On-Off switch to run the simulation. Press the space bar on the keyboard to make the CLK input rise from zero (0) to one (1) to produce a positive edge, then press the space bar again to bring the clock pulse back down (0).

Question: Did the output (Q) set (1) or clear (0)? On the positive clock edge or the negative clock edge?

Step 3.   Press the S and R keys to change inputs S and R so that they are the inverse of the previous values (S = 0 and R = 1). Press the space bar on the keyboard to make the CLK input rise from zero (0) to one (1) to produce a positive edge, then press the space bar again to bring the clock pulse down (0).

Questions: Did the output (Q) set (1) or clear (0)? On the positive clock edge or the negative clock edge?

Is this a positive or a negative edge-triggered flip-flop?

Step 4.   Set both the S and R inputs to binary zero (0). Keep pressing the space bar on the keyboard to produce a series of positive and negative clock edges, then stop the simulation.

Question: What happened to the flip-flop output (Q)? **Explain.**

Step 5.   Open circuit file FIG22-2. You are looking at an S-R latch wired as a positive-edge-triggered D flip-flop. See the Theory section for more details about the operation of this circuit.

Step 6.   Both the D and the CLOCK switches should be down (0). Click the On-Off switch to run the simulation. Press the space bar on the computer keyboard to make the clock input rise from zero (0) to one (1) to produce a positive edge, then press the space bar again to produce a negative edge.

Questions: Did the output (Q) set (1) or clear (0)? On the positive or negative clock edge? **Explain why.**

Step 7.   Press the D key on the keyboard to make input D be the inverse of the previous value.

Question: Did the output (Q) change state? **Explain.**

Step 8.   Press the space bar on the keyboard again to make the CLK input rise from zero (0) to one (1) to produce a positive edge, then press the space bar again to produce a negative edge. Now stop the simulation.

Questions: Did the output (Q) set (1) or clear (0)? On the positive clock edge or negative clock edge? **Explain why.**

Is this a positive or a negative edge-triggered flip-flop?

Step 9.   Open circuit file FIG22-3. You will test a 7474 positive-edge-triggered D flip-flop with active low asynchronous preset (PR) and clear (CLR) inputs. See the Theory section for a detailed discussion of the 7474 flip-flop.

Step 10.   The D and CLOCK switches should be down (0). The S (set) and C (clear) switches should be up (1). Click the On-Off switch to run the simulation. Repeat Steps 6–8 for the circuit in Figure 22-4.

NOTE: When the D flip-flop is first turned on, it will either set (1) or clear (0), depending on the internal transients.

Question: Were the results the same as in Steps 6–8?

No CHANGES wiTH Clock ⎰ AS Long AS S ANd/oR C ARE oN
No CHANGES wiTH D ⎱

It CHANGES wiTH Clock AS Long AS D HAS CHANGED FRom PREVIOUS CYCLE.

Step 11.   If the output is clear (Q = 0), press the S key on the computer keyboard to drop the active
            low preset (PR) input to binary zero (0). If the output is set (Q = 1), press the C key to drop
            the active low clear (CLR) input to binary zero (0).

Question: Did the output (Q) change state without a positive clock edge being applied? **Explain.**

> YES. BECAUSE PR AND CLR HAVE PRIORITY OVER CLOCK

Step 12.   Press the S or C key on the keyboard so both switches are up (1).

Question: Did the output (Q) change state? **Explain.**

> No. It's a DOUBLE JAM, SHOULDN'T BE BOTH UP.

Step 13.   Repeat Steps 11 and 12, then stop the simulation.

Questions: Did the output (Q) change state without a positive clock edge being applied?

> No.

Are the 7474 preset (PR) and clear (CLR) inputs asynchronous or synchronous inputs?

> ASYNCHRONOUS.

Step 14.   Open circuit file FIG 22-4. You are looking at a circuit that will display the timing of the
            7474 positive-edge-triggered D flip-flop. Bring down the word generator enlargement and
            make sure that the following settings are selected: Frequency = 1 kHz, Trigger = Internal,
            Controls = Burst, Hex codes (00000000, 00000001, 00000000, 00000001, 00000002,
            00000003, repeated twice, then 00000000, 00000001, repeated twice). Bring down the logic
            analyzer enlargement and make sure that the following settings are selected: Clocks/Div =
            16, Clock Setup (Clock Source = Internal, Clock Rate = 10 kHz, Pre-trigger Samples = 100,
            Post-trigger Samples = 1000, Threshold Voltage = 2.5), Trigger Settings (Trigger Clock
            Edge = Positive, Trigger Qualifier = x, Pattern A= xxxxxxxxxxxxxxxx, Trigger
            Combinations = A).

NOTE: If this experiment is being performed in a hardwired laboratory, use two pulse generators
in place of the word generator and a dual trace oscilloscope in place of the logic analyzer if a logic
analyzer is not available.

Step 15.   Click the On-Off switch to run the simulation. The word generator is applying a series of clock pulses to the clock input (CLK) and a pulse waveshape to the D input of the flip-flop. The logic analyzer is monitoring the flip-flop CLK input (green), the D input (red), output Q (blue), and inverted output Q′ (brown). After a full screen display, stop the simulation.

Questions: Does the flip-flop output Q (blue) change state on positive or negative clock edges?

POSITIVE

What determines whether the flip-flop output Q (blue) sets (Q = 1) or clears (Q = 0)?

BOTH DATA AND CLOCK.

Step 16.   Open circuit file FIG22-5. You will test a 74112 negative-edge-triggered J-K flip-flop with active low asynchronous preset (PR) and clear (CLR) inputs. See the Theory section for a detailed discussion of the 74112 J-K flip-flop.

Step 17.   The J switch should be down (0) and the K switch should be up (1). The CLOCK, S (set) and C (clear) switches should be up (1). Click the On-Off switch to run the simulation. Press the space bar on the keyboard to make the CLK input drop from one (1) to zero (0) to produce a negative edge, than press the space bar again to produce a positive edge.

Questions: Did the output (Q) set (1) or clear (0)? On the negative edge or the positive edge of the clock pulse?

CLEAR.   POSITIVE

Step 18.   Change inputs J and K so that they are the inverse of the previous values. Press the space bar on the keyboard to make the CLK input fall from one (1) to zero (0) to produce a negative edge, then press the space bar again to produce a positive edge.

Questions: Did the output (Q) set (1) or clear (0)? On the negative or positive edge of the clock pulse?

SET.   POSITIVE

Is this a positive or negative edge-triggered flip-flop?

POSITIVE EDGE

Which is the set input and which is the clear input (J or K) on a J-K flip-flop?

J = SET

K = CLEAR

Step 19.   Set both the J and K inputs to binary zero (0). Keep pressing the space bar on the keyboard
to produce a series of negative and positive clock edges.

Question: What happened to the flip-flop output (Q)? **Explain.**

Nothing. It needs J or K to be 1 to change

Step 20.   Set both the J and K inputs to binary one (1). Keep pressing the space bar on the keyboard
to produce a series of negative and positive clock edges.

Question: What happened to the flip-flop output (Q)? **Explain why.**

It changed. It needs J or K to be 1 to change.

Step 21.   Keep pressing the space bar on the keyboard until the output is clear (Q = 0). Now press the
S (set) key to drop the 74112 preset (PR) input to binary zero (0).

Questions: What happened to the flip-flop output (Q)?

Q = 1

Did the output change without a clock pulse being applied? **Explain why.**

Yes. S and C have priority

Step 22.   Press the S (set) key to raise the 74112 preset (PR) input to binary one (1). Press the C
(clear) key to drop the 74112 clear (CLR) input to binary zero (0), then stop the simulation.

Questions: What happened to the flip-flop output (Q)? **Explain why.**

Q = 0. S and C have priority

Did the output change without a clock pulse being applied?

Yes

Are the 74112 preset (PR) and clear (CLR) inputs asynchronous or synchronous inputs? **Explain why.**

ASYNCHRONOUS

Step 23.   Open circuit file Fig 22-6. You are looking at a circuit that will display the timing of the 74112 negative-edge-triggered J-K flip-flop. Bring down the word generator enlargement and make sure that the following settings are selected: Frequency = 1 kHz, Trigger = Internal, Controls = Burst, Hex codes (00000007, 00000006, 00000005, 00000004, 00000003, 00000002, 00000001, 00000000, repeated two times). Bring down the logic analyzer enlargement and make sure that the following settings are selected: Clocks/Div = 16, Clock Setup (Clock Source = Internal, Clock Rate = 10 kHz, Pre-trigger Samples = 100, Post-trigger Samples = 1000, Threshold Voltage = 2.5), Trigger Settings (Trigger Clock Edge = Positive, Trigger Qualifier = x, Pattern A= xxxxxxxxxxxxxxxx, Trigger Combinations = A).

NOTE: If this experiment is being performed in a hardwired laboratory, use three pulse generators in place of the word generator and a dual trace oscilloscope in place of the logic analyzer if a logic analyzer is not available.

Step 24.   Click the On-Off switch to run the simulation. The word generator is applying a series of clock pulses to the clock input (CLK), a pulse waveshape to the J input, and a different pulse waveshape to the K input of the flip-flop. The logic analyzer is monitoring the CLK input (green), the J input (red), the K input (brown), output Q (blue), and inverted output Q' (orange). After a full screen display, stop the simulation.

Questions: Does the flip-flop output Q (blue) change state on the positive or negative clock edges?

NEGATIVE.

Does the flip-flop output Q (blue) toggle when the J and K inputs are both binary one (1) during a negative clock edge?

Yes

Does the flip-flop output (Q) (blue) set (Q = 1) when J = 1 and K = 0 during a negative clock edge?

Yes

Does the flip-flop output Q (blue) clear (Q = 0) when J = 0 and K = 1 during a negative clock edge?

yes

Does the flip-flop output Q (blue) stay the same when J = 0 and K = 0 during a negative clock edge?

Yes

EXPERIMENT

23 Monostable and Astable Multivibrators

Objectives:

1.   Investigate the operation of a monostable multivibrator (one-shot)
2.   Investigate the operation of a 555 timer wired as an astable multivibrator.

Materials:

One 5 V dc power supply
One monostable multivibrator (one shot) (1-74121 IC)
One 555 timer (LM555 CN)
One square wave generator
One dual-trace oscilloscope
Capacitors—0.02 μF, 0.01 μF(2), 1 μF
Resistors—72 kΩ, 48 kΩ (2), 10 kΩ, 5 kΩ, 1 kΩ

Theory:

Latches and flip-flops, which were studied in the previous experiments, are often referred to as **bistable multivibrators** because the output has **two stable states** (high or low). **A monostable multivibrator,** otherwise known as a **one-shot**, has only **one stable output state**. When it is triggered, the output will go to the **unstable state (Q = 1)** for a predetermined period of time, and then return to the **stable state (Q = 0)**. The time period that the output will stay in the unstable state (Q = 1) is determined by the RC circuit connected to the one-shot. A **nonretriggerable one-shot** will not respond to any additional trigger pulses during the time that it is in the unstable state (Q = 1). Therefore, the time that the output is unstable (Q = 1) for a nonretriggerable one-shot will always be dependent on the RC time constant only. A **retriggerable one-shot** will respond to trigger pulses during the time that the output is in the unstable state (Q = 1). Therefore, the unstable state (Q = 1) for a retriggerable one-shot will restart on each trigger pulse, causing the time that the output is unstable (Q = 1) to depend on the trigger pulse timing.

An **astable multivibrator** doesn't have any stable output state. Therefore it continuously switches between two unstable states without any triggering. This results in a **square wave output** that is useful for providing clock pulses for synchronous digital circuits. An astable multivibrator is often referred to as a **free-running pulse oscillator or generator**.

The circuit in Figure 23-1 will be used to demonstrate the operation of a **monostable multivibrator (one-shot).** The pulse generator will apply a series of short-duration negative trigger pulses to the negative-edge-triggered trigger input (A2) to trigger the one-shot. The time period (t_w) that the one-shot will stay in the unstable state ($Q = 1$) is determined by the value of resistor R and capacitor C using the equation

$$t_w = 0.7(R)(C)$$

The oscilloscope will monitor the one-shot output (Q) and trigger input (A2).

The 555 timer in Figure 23-2 is wired as an astable multivibrator (free-running pulse generator). The external components RA, RB, and C determine the frequency of oscillation (f) and the duty cycle. The control input (CON) is not being used. The $0.01\,\mu F$ capacitor connected to the control input (CON) is strictly for decoupling the control input (CON) from noise pickup and has no effect on the oscillating frequency. The **frequency of oscillation (f)** is the inverse of the **time period for one pulse cycle (T).** Therefore,

$$f = \frac{1}{T}$$

The **duty cycle** is the time that the output is high (t_2) divided by the total time period for one pulse cycle (T), expressed as a percentage. Therefore,

$$\text{Duty cycle} = \frac{t_2}{T} \times 100\%$$

The **expected frequency of oscillation (f)** and the **expected duty cycle** can be calculated from the values of RA, RB, and C by first calculating the expected time that the output is low (t_1), the expected time that the output is high (t_2), and the expected time period for one pulse cycle (T) from the equations

$$t_1 = 0.7(RB)(C)$$
$$t_2 = 0.7(RA + RB)(C)$$
$$T = t_1 + t_2$$

For obtaining a 50% duty cycle,

$$\frac{t_2}{T} = 0.5$$

Therefore,

$$t_2 = 0.5(T) = \frac{T}{2}$$

This can only be achieved if RB is much larger than RA, (RA cannot be zero) so that

$$t_2 = 0.7(RA + RB)(C) \approx 0.7(RB)(C) = t_1$$

Therefore,

$$T = t_1 + t_2 \approx t_2 + t_2 = 2(t_2)$$

and $t_2 \approx \dfrac{T}{2}$

Figure 23-1 Monostable Multivibrator (One-Shot)

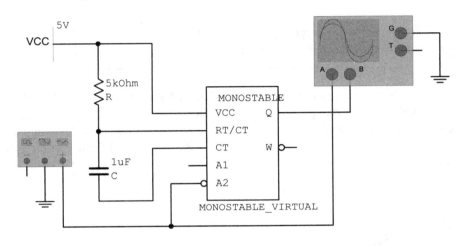

Figure 23-2 555 Astable Multivibrator

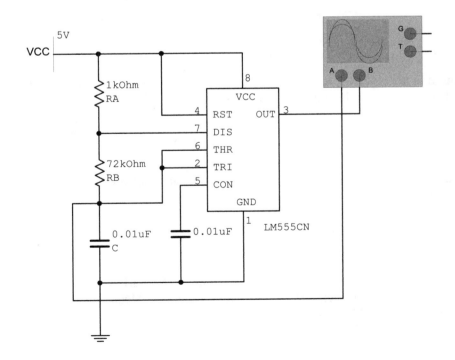

Procedure:

Step 1.   Open circuit file FIG 23-1. You are looking at a monostable multivibrator (one-shot) connected to a pulse generator and oscilloscope. Bring down the function generator enlargement and make sure that the following settings are selected: *Square Wave*, Freq = 200 Hz, Duty Cycle = 95%, Ampl = 2.5 V, Offset = 2.5 V. Move the function generator to the left. Bring down the oscilloscope enlargement and make sure that the following settings are selected: Time base (Scale = 2 ms/Div, Xpos = 0, Y/T), Ch A (Scale = 5 V/Div, Ypos = –2, DC), Ch B (Scale = 5 V/Div, Ypos = 0, DC), Trigger (Neg edge, Level = 0, Sing, A). Move the oscilloscope to the right.

> NOTE: If this experiment is performed in a hardwired laboratory, use a 74121 monostable multivibrator (one-shot) and connect trigger input B and trigger input A1 to +5 V (1).

Step 2.   Click the On-Off switch to run the simulation. After a full screen display, stop the simulation. The pulse generator is applying a series of short duration trigger pulses to the one-shot negative-edge-triggered trigger input (A2). The oscilloscope is monitoring the one-shot trigger input, A2 (red curve plot), and the one-shot output, Q (blue curve plot). Measure and record the high level (unstable state) pulse width (t_w) at the one-shot output (blue curve plot).

$t_w = \underline{\hspace{2in}}$

Question: Is the one-shot being triggered on the positive edge or the negative edge of the trigger input pulse (red curve plot)? **Explain how you determined the answer.**

Step 3.   Based on the value of resistor R and capacitor C, calculate the expected unstable (Q = 1) pulse width (t_w) at the output (Q) of the one-shot.

Question: How did your calculated value for the output unstable pulse width (t_w) compare with the measured value in Step 2?

Step 4.   Change the value of resistor R to 10 kΩ. Click the On-Off switch to run the simulation again. After a full screen display, stop the simulation. Measure and record the time between input pulses (T).

$T = \underline{\hspace{2in}}$

Step 5.   With the new value of R, calculate the new expected unstable (Q = 1) pulse width (t_w).

Questions: Based on the calculated unstable pulse width (t_w) for the one-shot output (blue) and the time between input pulses (T) for the trigger input (red), is this a **nonretriggerable** or **retriggerable** monostable multivibrator? **Explain how you determined the answer.**

Step 6.   Open circuit file FIG23-2. You are looking at a 555 timer wired as an astable multivibrator (free-running pulse generator). Bring down the oscilloscope enlargement and make sure that the following settings are selected: Time base (Scale = 500 µs/Div, Xpos = 0, Y/T), Ch A (Scale = 2 V/Div, Ypos = 0, DC), Ch B (Scale = 2 V/Div, Ypos = 0, DC), Trigger (Neg edge, Level = 0, Sing, B). Move the oscilloscope to the right. Click the On-Off switch to run the simulation. After a full screen display, stop the simulation.

Step 7.   Measure and record the time that the output is low (t_1), the time that the output is high (t_2), and the total time period for one cycle (T) of the output (blue curve plot).

            $t_1 = $ _____ $t_2 = $ _____ $T = $ _____

Step 8.   Measure and record the trigger voltage (red) that switches the output high (V_{high}) and the trigger voltage (red) that switches the output low (V_{low}).

            $V_{high} = $ _____ $V_{low} = $ _____

Questions: What percentage of Vcc is V_{high}?

What percentage of Vcc is V_{low}?

Step 9.   Based on the time period (T) measured in Step 7, calculate the pulse frequency (f) in hertz.

Step 10.   Based on t_1, t_2, and T measured in Step 7, calculate the duty cycle.

Step 11.   Based on the values of RA, RB, and C for the circuit in Figure 23-2, calculate the expected
             values of t_1, t_2, and T.

Question: How did your new calculated values for t_1, t_2, and T compare with the measured values in
Step 7?

Step 12.   Change resistor RA and resistor RB to 48 kΩ. Click the On-Off switch to run the
             simulation. After a full screen display, stop the simulation.

Step 13.   Measure the record the time that the output is low (t_1), the time that the output is high (t_2)
             and the total time period for one cycle (T) of the output (blue curve plot).

             $t_1 = $ _____ $t_2 = $ _____ $T = $ _____

Step 14.   Based on the time period (T) measured in Step 13, calculate the new pulse frequency (f) in
             hertz.

Step 15.   Based on t_1, t_2, and T measured in Step 13, calculate the new duty cycle.

Question: How does the new duty cycle compare with the duty cycle calculated in Step 10? **Explain the reason for any difference.**

Step 16.   Based on the new values of RA and RB, and the value of C, for the circuit in Figure 23-2, calculate the new expected values of t_1, t_2, and T.

Questions: How did your new calculated values for t_1, t_2, and T compare with the measured values in Step 13?

How did the new values of t_1, t_2, and T compare with the values in Steps 7 and 11? **Explain the reason for any difference.**

Step 17.   Change capacitor C to 0.02 μF. Click the On-Off switch to run the simulation. After a full screen display, stop the simulation.

Step 18.   Measure and record the time that the output is low (t_1), the time that the output is high (t_2), and the total time period for one cycle (T) of the output (blue curve plot).

$t_1 = $ _____ $t_2 = $ _____ $T = $ _____

Question: How did the values for t_1, t_2, and T compare with the measured values in Step 13 with $C = 0.01$ μF? **Explain the reason for any difference.**

Step 19.   Based on the values of t_1, t_2, and T, calculate the new duty cycle.

Question: How does the new duty cycle compare with the duty cycle calculated in Step 15 with C = 0.01 μF? **Explain your answer.**

Step 20.   Based on the new time period (T), calculate the new pulse frequency (f) in hertz.

Question: How did the new frequency compare with the frequency determined in Step 14 with C = 0.01 μF?

Name_____

Date_____

24 Registers and Data Storage

Objectives:

1. Demonstrate the application of flip-flops in serial and parallel registers.
2. Demonstrate parallel register data transfer.
3. Demonstrate serial register data transfer.
4. Investigate the operation of a 74173 four-bit parallel in/parallel out register
5. Investigate the operation of a 74194 universal four-bit bidirectional shift register.
6. Observe shift register timing.

Materials:

One 5 V dc power supply
Nine logic switches
Nine logic probe lights
Four positive-edge-triggered D-type flip-flops (2-7474 ICs)
Seven two-input AND gates (2-7408 ICs)
Three two-input OR gates (1-7432 IC)
One INVERTER (1-7404 IC)
One four-bit parallel in/parallel out register (1-74173 IC)
One universal four-bit bidirectional shift register (1-74194 IC)
Two pulse generators
One logic analyzer or dual-trace oscilloscope

Theory:

A **register** consists of a series of flip-flops for the purpose of **storing and transferring binary data** in a digital system. Each register flip-flop stores one bit of binary data. The storage capacity of the register is determined by the number of flip-flops, which determines the number of binary bits it can store. Registers can be **loaded** in **serial** one bit at a time, or in **parallel** with all bits loaded at one time. Binary data can be **output** from a register in **serial** one bit at a time, or in **parallel** with all bits output at one time. Register input and output is controlled by the positive or negative edge of a clock pulse applied to the register clock terminal. Some registers can also be **input enabled or disabled** and **output enabled or disabled**.

The circuit in Figure 24-1 consists of four **positive-edge-triggered D-type flip-flops** wired as a **4-bit parallel in/parallel out register**. The 4-bit input consists of D3, D2, D1, and D0, and the 4-bit output consists of Q3, Q2, Q1, and Q0. Each flip-flop stores one bit of binary data. Because all four flip-flop clock inputs are connected together, they receive the positive clock edge at the same time. This causes the bits to be stored simultaneously in each flip-flop, making this a parallel in register. Because all four flip-flop outputs are being monitored simultaneously, this is also a parallel out register. This register can be cleared by clearing each flip-flop (Q = 0). This is accomplished by connecting the **active low asynchronous clear inputs** of the flip-flops together to form an **active low CLR′** terminal. When a binary zero (0) is applied to the CLR′ terminal, the register will clear (Q = 0). The CLR′ terminal must be brought back to binary one (1) if you want to store new binary data in the register, otherwise the register will stay cleared because the flip-flop asynchronous clear inputs **override the clock inputs**.

The circuit in Figure 24-2 consists of four **positive-edge-triggered D-type flip-flops** wired as a **4-bit serial in/parallel out register**. Each flip-flop stores one bit of binary data. Because all four flip-flop clock inputs are connected together, they receive the positive clock edge at the same time. Data is input to this register into the first flip-flop one bit at a time at input Din. Because each flip-flop output is connected to the next flip-flop input, each data bit is **shifted right** on each clock pulse **positive edge**, making this a **serial in shift register**. Therefore, it takes **four clock pulses** to store 4 bits of binary data in this register. Because all four flip-flop outputs (Q3, Q2, Q1, and Q0) are being monitored simultaneously, this is a **parallel out register**.

The circuit in Figure 24-3 consists of four **positive-edge-triggered D-type flip-flops** wired as a **4-bit serial in/serial out register**. Each flip-flop stores one bit of binary data. Because all four flip-flop clock inputs are connected together, they receive the positive clock edge at the same time. Data is input to this register into the first flip-flop one bit at a time at input Din. Because each flip-flop output is connected to the next flip-flop input, each data bit is **shifted right** on each clock pulse **positive edge**, making this a **serial in shift register**. Therefore, it takes **four clock pulses** to store 4 bits of binary data in this register. Because only the last flip-flop output is being monitored, this register must be clocked **three times** to shift all four bits to the output (Q). Therefore, this is a **serial out register**. Remember, the last flip-flop output can be read before any clock pulses are applied. This is the reason that only three clock pulses are needed to read the 4-bit output data.

The circuit in Figure 24-4 consists of four **positive-edge-triggered D-type flip-flops** and some additional logic circuitry wired as a **4-bit parallel in/serial out register**. The **S/L′ input** is the **shift/load** input. A **binary zero (0)** on this input will cause the logic circuitry to connect each input bit (D3–D0) to the D input on each flip-flop. Therefore, this register will **parallel load** the input data (D3–D0) on the **positive edge** of the clock pulse when the S/L′ input is at binary zero (0), making this a **parallel in register**. A **binary one (1)** on the S/L′ input will cause the logic circuitry to shift each stored bit to the next flip-flop on each positive edge of the clock pulse. Because only the last flip-flop is being monitored, a binary one (1) on the S/L′ input makes it possible to read the register output (Q) one bit at a time, making this a **serial out register**.

The 74173 in Figure 24-5 is a **4-bit parallel in/parallel out register.** This register has two **active low input enables (G1 and G2)** and two **active low output enables (M and N)**. It also has an **active high clear (CLR)** and a **positive-edge-triggered clock input (CLK)**. The 74173 internal circuitry is similar to the circuitry in Figure 24-1 with some additional logic circuitry to control the input enable and output enable.

The 74194 in Figure 24-6 is a **universal 4-bit bidirectional shift register**. This register has a **parallel load data input (D–A)** and two **serial load data inputs (SL and SR)**. Serial load data input SL is a **shift left** serial input, and serial load data input SR is a **shift right** serial input. **Shift left** means shifting from **QD to QA (Q3 to Q0)** and **shift right** means shifting from **QA to QD (Q0 to Q3)**. The **mode** control inputs (S1 S0) determine whether the register shifts left, shifts right, parallel loads, or stays the same when a positive clock edge is applied to the CLK input. See Table 24-1 for the codes.

Table 24-1 74194 Mode Control Inputs

S1 S0	Result on CLK Edge
0 0	No change
0 1	Shift right (Q0 to Q3)
1 0	Shift left (Q3 to Q0)
1 1	Parallel load

The circuit in Figure 24-7 will display the **timing** of a 74194 shift register. The **mode control input (S1 S0)** determines the register mode. (See Table 24-1). The word generator will apply a series of clock pulses to the register **clock input (CLK)** and a pulse waveform to the **serial data inputs (SL and SR)**. The logic analyzer will monitor the clock input (CLK), the serial data inputs (SL and SR), and the four outputs (QD–QA).

Figure 24-1 Parallel In/Parallel Out Register

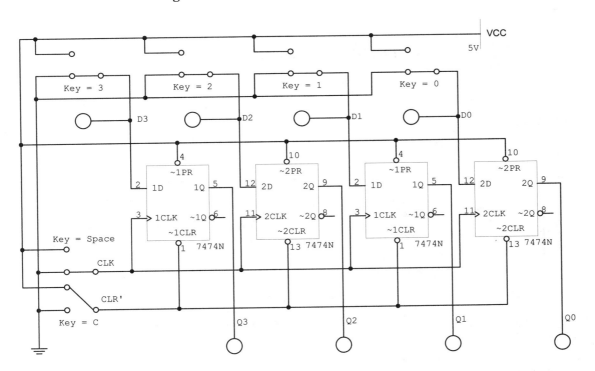

Figure 24-2 Serial In/Parallel Out Register

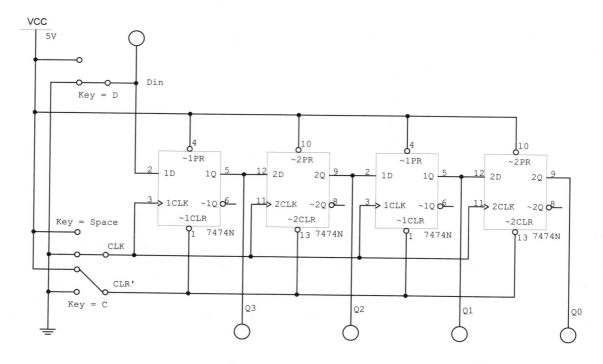

Figure 24-3 Serial In/Serial Out Register

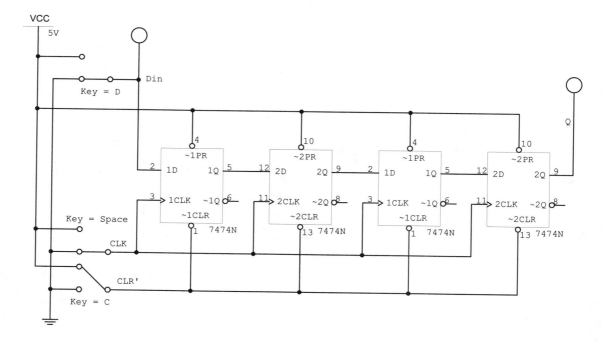

Figure 24-4 Parallel In/Serial Out Register

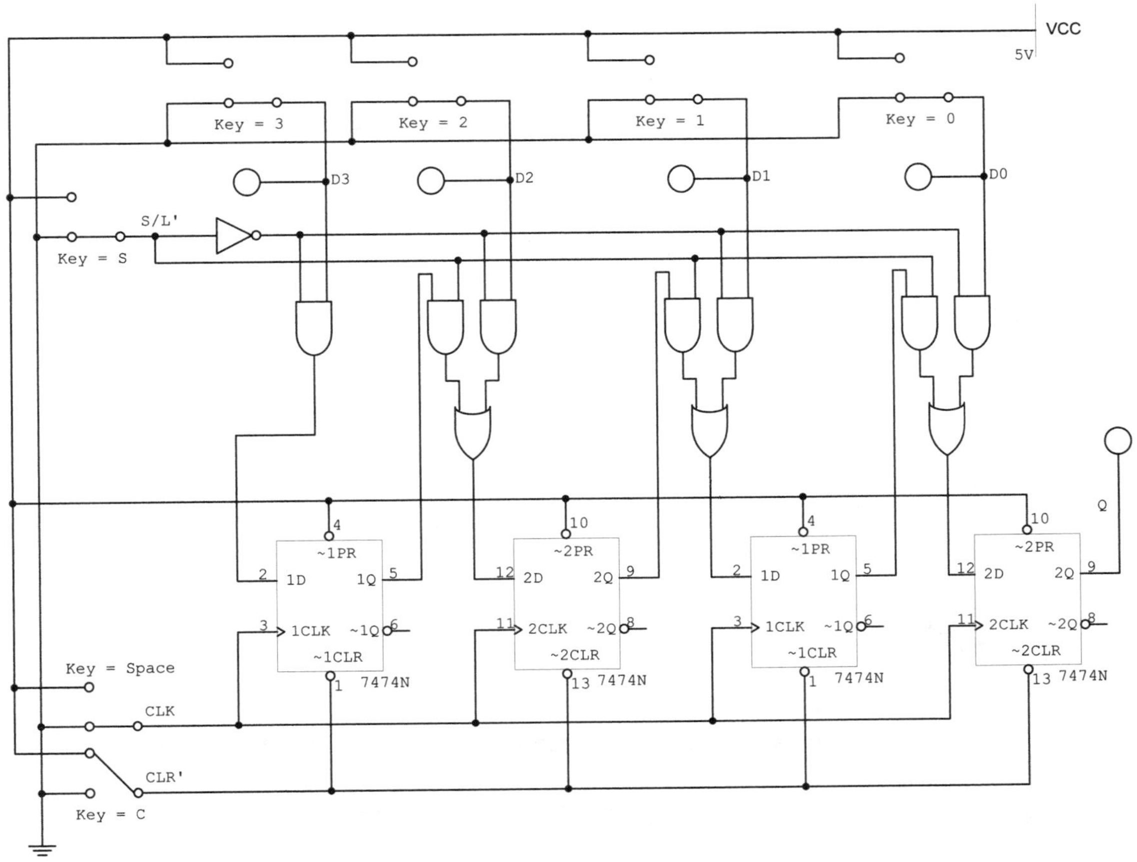

Figure 24-5 74173 Four-Bit Parallel In/Parallel Out Register

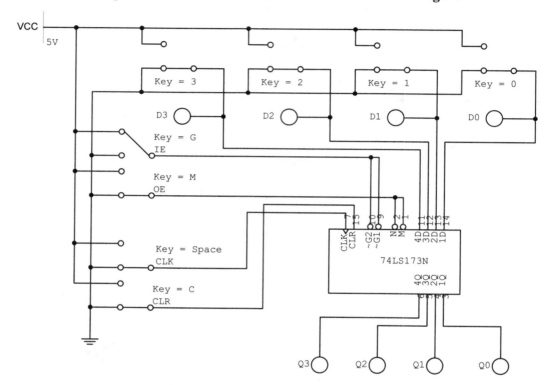

Figure 24-6 74194 Universal Four-Bit Bidirectional Shift Register

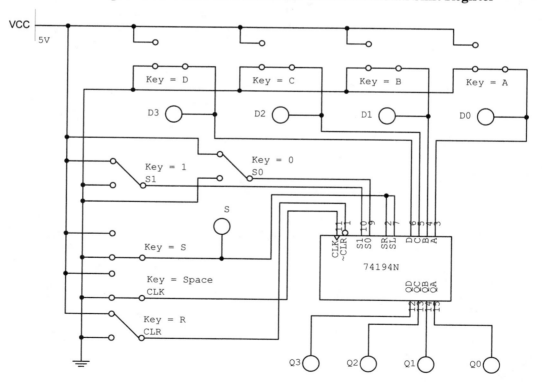

Figure 24-7 74194 Shift Register Timing

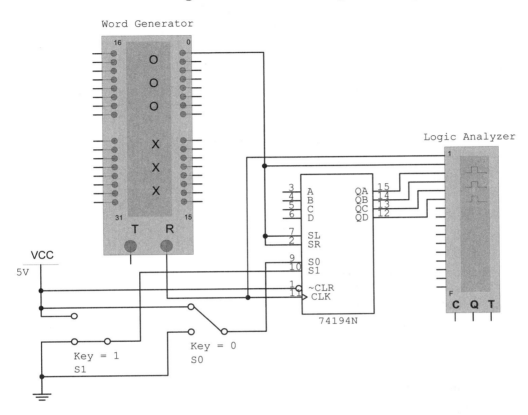

Procedure:

Step 1.   Open circuit file FIG24-1. You are looking at four positive-edge-triggered D-type flip-flops wired as a 4-bit parallel in/parallel out register. The CLR switch should be up (1) and the CLK switch should be down (0). Click the On-Off switch to run the simulation and press the C key on the computer keyboard to clear the register, if needed. Press the C key again to raise the CLR switch back to binary one (1).

NOTE: If the switch doesn't change, click the mouse arrow in the circuit window before pressing C.

Questions: Is the clear input (CLR') active low or active high?

Why does the CLR switch need to be up (1) to store data in the register?

Step 2.   Set the data input switches (0–3) to any 4-bit binary number other than zero.

Question: Did the register output (Q3–Q0) change? **Explain.**

Step 3.   Press the space bar on the computer keyboard to make the CLK input rise from zero (0) to one (1) to produce a positive clock edge.

Questions: What did you observe at the register output (Q3–Q0)? **Explain.**

Is this register positive-edge-triggered or negative-edge-triggered?

Did the input data load in parallel or serial?

Is the data being read at the output in parallel or serial?

How many clock pulses were required to store the 4-bit binary input data?

Step 4.   Change the binary data input (D3–D0) by changing the data switches (0–3). Press the space bar on the computer keyboard to make the CLK input drop from one (1) to zero (0) to produce a negative clock edge.

Question: Did the register output (Q3–Q0) change? **Explain.**

Step 5.   Press the space bar on the computer keyboard again to make the CLK input rise from zero (0) to one (1) to produce a positive clock edge and observe the outputs, then stop the simulation.

Question: What did you observe at the register output (Q3–Q0)? **Explain.**

Step 6.   Open circuit file FIG24-2. You are looking at four positive-edge-triggered D-type flip-flops
             wired as a 4-bit serial in/parallel out register. The CLR switch should be up (1) and the
             CLK switch should be down (0). Click the On-Off switch to run the simulation and press
             the C key on the computer keyboard to clear the register, if needed. Press the C key again
             to raise the CLR switch back to binary one (1).

Step 7.   Press the D key on the keyboard to set the data input switch (D) to binary one (up) so that
             Din = 1.

Question: Did the register output (Q3–Q0) change? **Explain.**

Step 8.   Press the space bar on the computer keyboard to make the CLK input rise from zero (0) to
             one (1) to produce a positive clock edge.

Question: What did you observe at the register output (Q3–Q0)? **Explain.**

Step 9.   Press the D key on the keyboard again to set the data input switch (D) to binary zero
             (down) so that Din = 0. Press the space bar on the computer keyboard to make the CLK
             input fall from one (1) to zero (0) to produce a negative clock edge, and press the space bar
             again to produce a positive clock edge.

Question: What did you observe at the register output (Q3–Q0)? **Explain.**

Step 10.   Repeat Steps 7–9, then stop the simulation.

Questions: What did you observe at the register output (Q3–Q0)?

Is this register positive-edge-triggered or negative-edge-triggered?

Did the input data load in parallel or serial?

Is the data being read at the output in parallel or serial?

How many clock pulses were required to store the 4-bit binary input data?

Step 11.   Open circuit file FIG24-3. You are looking at four positive-edge-triggered D-type flip-flops wired as a 4-bit serial in/serial out register. The CLR switch should be up (1) and the CLK switch should be down (0). Click the On-Off switch to run the simulation and press the C key on the computer keyboard to clear the register. Press the C key again to raise the CLR switch back to binary one (1).

Step 12.   Press the D key on the keyboard to set the data input switch (D) to binary one (up) so that Din = 1. Press the space bar on the computer keyboard to make the CLK input rise from zero (0) to one (1) to produce a positive clock edge.

Step 13.   Press the D key on the keyboard to set the data input switch (D) to binary zero (down) so that Din = 0. Press the space bar on the computer keyboard to make the CLK input fall from one (1) to zero (0) to produce a negative clock edge, and press the space bar again to produce a positive clock edge.

Step 14.   Repeat Steps 12 and 13.

Question: What did you observe at the register output (Q) after the fourth positive clock edge? **Explain.**

Step 15.   Press the space bar on the computer keyboard enough times to produce three positive clock edges and observe the register output (Q), then stop the simulation.

Questions: What did you observe at the register output (Q) on each positive edge of the clock?

Is this register positive-edge-triggered or negative-edge-triggered?

Did the input data load in parallel or serial?

Is the data being read at the output in parallel or serial?

How many clock pulses were required to store the 4-bit binary input data?

How many clock pulses were required to read the 4-bit binary output data?

Step 16. Open circuit file FIG24-4 and click "Zoom Out" to see the entire circuit. You are looking at four positive-edge-triggered D-type flip-flops and some additional logic circuitry wired as a 4-bit parallel in/serial out register. The S/L' input is the shift/load input. A binary zero (0) on this input will cause this register to parallel load the input data (D3–D0) on the positive edge of the clock pulse. A binary one (1) on this input will cause the register to shift each stored bit to the output (Q) one bit at a time on each positive edge of the clock pulse. The C switch should be up (1) and the CLK switch and the S switch should be down (0).

NOTE: If this experiment is being performed in a hardwired laboratory, use two 7474 ICs for the positive-edge-triggered D-type flip-flops, two 7408 ICs for the 2-input AND gates, one 7432 IC for the 2-input OR gates, and one 7404 IC for the INVERTER.

Step 17. Click the On-Off switch to run the simulation and press the C key on the computer keyboard to clear the register. Press the C key again to raise the CLR switch back to binary one (1).

Step 18. Set the data input switches (0–3) to any 4-bit binary number other than zero. Make sure the S switch is down (0). Press the space bar on the computer keyboard to make the CLK input rise from zero (0) to one (1) to produce a positive clock edge. The binary input data (D3–D0) should have been parallel loaded into the register on the positive edge of the clock pulse even though all of the outputs are not being monitored.

Step 19. Press the S key on the computer keyboard to set the S/L' input to binary one (1). Press the space bar on the computer keyboard enough times to produce three positive clock edges and observe the register output (Q) after each pulse, then stop the simulation.

Questions: What did you observe at the register output (Q) on each positive edge of the clock pulse?

Is this register positive-edge-triggered or negative-edge-triggered?

Did the input data load in parallel or serial?

Is the data being read at the output in parallel or serial?

How many clock pulses were required to store the 4-bit binary input data?

How many clock pulses were required to read the 4-bit binary output data?

Step 20.   Open circuit file FIG24-5. You will test a 74173 4-bit parallel in/parallel out register. This register has two active low input enables (G1 and G2) which are activated by switch IE, and two active low output enables (M and N), which are activated by switch OE. It also has an active high clear (CLR) and a positive-edge-triggered clock (CLK) input. Switch IE should be up (1) and switches OE, CLK, and CLR should be down (0).

Step 21.   Click the On-Off switch to run the simulation. If the register output is not cleared, press the C key on the computer keyboard to raise the CLR switch to clear the register. Press the C key again to lower the CLR switch back to binary zero (0).

Step 22.   Set the data input switches (0–3) to any 4-bit binary number other than zero. Press the G key on the computer keyboard to lower the IE switch (active low input enable) to zero (0) to enable the input (D).

Question: Did the register output (Q3–Q0) change? **Explain.**

Step 23.   Press the space bar on the computer keyboard to make the CLK input rise from zero (0) to one (1) to produce a positive clock edge.

Questions: What did you observe at the register output (Q3–Q0)?

Is this register positive-edge-triggered or negative-edge-triggered?

Did the input data load in parallel or serial?

Is the data being read at the output in parallel or serial?

How many clock pulses were required to store the 4-bit binary input data?

Step 24.   Press the M key on the computer keyboard to raise the OE switch (active low output enable) to binary one (1).

Question: What did you observe at the register output (Q3–Q0)? **Explain.**

Step 25.   Press the M key again to lower switch OE (active low output enable) to zero (0). Press the G key to raise switch IE (active low input enable) to one (1) to disable the input (D). Change the data input switches (0–3) to a new 4-bit binary number. Press the space bar on the computer keyboard enough times to make the CLK input rise from zero (0) to one (1) and produce a positive clock edge and observe the outputs (Q3–Q0), then stop the simulation.

Question: What did you observe at the register output (Q3–Q0)? **Explain.**

Step 26.   Open circuit file FIG24-6. You will test a 74194 universal 4-bit bidirectional shift register. This register has a parallel load data input (D–A) and two serial load data inputs (SL and SR). Serial load data input SL is a shift left serial input, and serial load data input SR is a shift right serial input. The mode control inputs (S1 S0) determine whether the register shifts left, shifts right, parallel loads, or stays the same when a positive clock edge is applied to the positive-edge-triggered CLK input (See Table 24-1 in the Theory section). Switch S provides the serial data inputs (SL and SR), switches D–A provide the parallel data inputs (D3–D0), and switches S1 and S0 control the mode inputs (S1 S0). The CLK and S switches should be down (0) and the CLR, S1, and S0 switches should be up (1).

Step 27.   Click the On-Off switch to run the simulation. If the register output is not cleared, press the R key on the computer keyboard to clear the register. Press the R key again to raise the CLR switch back to binary one (1).

Step 28.   With switches S1 and S0 up (S1 S0 = 11), set the data input switches (D–A) to any 4-bit binary number other than zero.

Question: Did the register output (Q3–Q0) change? **Explain.**

Step 29.   Press the space bar on the computer keyboard to make the CLK input rise from zero (0) to one (1) to produce a positive clock edge.

Questions: What did you observe at the register output (Q3–Q0)? **Explain.**

Is this register positive-edge-triggered or negative-edge-triggered?

Did the input data load in parallel or serial?

Is the data being read at the output in parallel or serial?

How many clock pulses were required to store the 4-bit binary input data?

Step 30.   Clear the register with the R key following the procedure in Step 27. Press the "1" key on the keyboard to set the mode control (S1 S0) to 01. Press the S key on the keyboard to raise the serial data input (S) to binary one (1). Press the space bar on the computer keyboard to make the CLK input fall from one (1) to zero (0) to produce a negative clock edge.

Questions: Did the register output (Q3–Q0) change?

Is this register positive-edge-triggered or negative-edge-triggered?

Step 31.   Press the space key on the computer keyboard to make the CLK input rise from zero (0) to one (1) to produce a positive clock edge.

Questions: What did you observe at the register output (Q3–Q0)? **Explain.**

Step 32.   Press the space bar on the computer keyboard enough times to produce three more positive clock edges and observe the register output (Q3–Q0).

Questions: What did you observe at the register output (Q3–Q0)? **Explain.**

Did the input data load in parallel or serial? If serial, is it shift left or shift right?

Is the data being read at the output in parallel or serial?

How many clock pulses were required to store the 4-bit binary input data?

Step 33.   Clear the register with the R key following the procedure in Step 27. Press the "1" key and the "0" key on the keyboard to set the mode control (S1 S0) to 10. Press the space bar on the computer keyboard enough times to produce four positive clock edges and observe the register output (Q3–Q0).

Questions: What did you observe at the register output (Q3–Q0)? **Explain.**

Did the input data load in parallel or serial? If serial, is it shift left or shift right?

Step 34.   Press the "1" key on the keyboard to set the mode control (S1 S0) to 00. Press the S key on the keyboard to set the serial input (S) to zero (0). Press the space bar on the computer keyboard enough times to produce four positive clock edges and observe the register output (Q3–Q0), then stop the simulation.

Question: What did you observe at the register output (Q3–Q0)? **Explain.**

Step 35.   Open circuit file FIG24-7. You are looking at a circuit that will display the timing of a 74194 4-bit shift register. The word generator is applying a series of clock pulses to the register clock input (CLK) and a pulse waveform to the serial data inputs (SL and SR). The logic analyzer is monitoring the clock input (CLK), the serial data inputs (SL and SR), and the four outputs (QD–QA). Switch S1 should be down (0) and switch S0 should be up (1). Bring down the word generator enlargement and make sure that the following settings are selected: Frequency = 1 kHz, Trigger = Internal, Controls = Burst, Hex codes (00000001, 00000000, 00000000, 00000000, repeated four times). Remove the word generator by clicking the x in the upper right corner of the generator. Bring down the logic analyzer enlargement and make sure that the following settings are selected: Clocks/Div = 16, Clock Setup (Clock Source = Internal, Clock Rate = 10 kHz, Pre-trigger Samples = 100, Post-trigger Samples = 1000, Threshold Voltage = 2.5), Trigger Settings (Trigger Clock Edge = Positive, Trigger Qualifier = x, Pattern A= xxxxxxxxxxxxxxxx, Trigger Combinations = A). Move the logic analyzer to the right so that switches S1 and S0 are visible. Click the On-Off switch to run the simulation. The light blue curve plot on the logic analyzer screen is the clock (CLK) input to the register. The red curve plot is the serial data input (SL and SR). The dark blue (QA), dark green (QB), brown (QC), and light green (QD) curve plots represent the 4-bit register output (QD–QA). After a full screen display, stop the simulation.

NOTE: If this experiment is being performed in a hardwired laboratory, use two pulse generators in place of the word generator. If a logic analyzer is not available, use a dual-trace oscilloscope.

Question: Did the input data load in parallel or serial? If serial, is it shift left or shift right?

Step 36.   Press the "1" and "0" keys on the keyboard to set the mode switches (S1 S0) to 10. Bring down the word generator enlargement and reset the cursor to the top of the column of numbers by right clicking the top number and selecting "Set Cursor." Remove the word generator by clicking the x. Click the On-Off switch to run the simulation again. After a full screen display, stop the simulation.

Question: Did the input data load in parallel or serial? If serial, is if shift left or shift right?

Name_____

Date_____

EXPERIMENT

25 | Asynchronous Counters

Objectives:

1. Demonstrate how J-K flip-flops are used to build an asynchronous (ripple) counter.
2. Demonstrate the effect of flip-flop propagation delay on asynchronous counter timing.
3. Demonstrate asynchronous (ripple) down counting.
4. Investigate the operation of a 74293 (7493) asynchronous (ripple) counter.
5. Demonstrate frequency division using asynchronous (ripple) counters.
6. Demonstrate how to change the modulus (divide-by) of an asynchronous (ripple) counter.

Materials:

One 5 V dc power supply
Four logic switches
Four logic probe lights
Four negative-edge-triggered J-K flip-flops (2-74112 ICs)
Two asynchronous (ripple) counters (2-74293 or 2-7493 ICs)
One function (pulse) generator
One logic analyzer or dual-trace oscilloscope
Four 1 kΩ resistors

Theory:

J-K flip flops in the **toggle mode** ($J = 1$, $K = 1$) can be connected together to build a **binary counter**. The number of flip-flops determines how high the counter will count and the number of binary states (counts). The **modulus (MOD)** of a counter is equal to the number of binary states.

Counters are classified into two basic categories based on the way they are clocked. In **asynchronous counters**, the first flip-flop receives the input clock pulse and each successive flip-flop is clocked by the output of the preceding flip-flop. This is the reason that asynchronous counters are often referred to as **ripple counters**. In **synchronous counters**, all flip-flops receive the input clock pulse simultaneously. Synchronous counters will be studied in the next experiment.

Flip-flop **propagation delay time (t_p)** is the time it takes for a flip-flop output to change state after the flip-flop clock input receives a clock edge. Because each flip-flop in an asynchronous (ripple) counter is not clocked until the preceding flip-flop output changes state, all of the outputs do not respond to the counter clock input at the same time. Therefore, the **time delay** between the counter clock input and the response of the last flip-flop output depends on the number (N) of counter flip-flops. If the clock frequency is too high, the last flip-flop output will not change state before the next counter input clock pulse is applied. This will cause errors in the count. For this reason, the **maximum clock frequency**

(f_{max}) for asynchronous (ripple) counters is lower than the maximum clock frequency for synchronous counters. For an asynchronous counter, the **maximum clock frequency (f_{max})** can be calculated from

$$f_{max} = \frac{1}{(N \times t_p)}$$

Asynchronous (ripple) counters are often used as **frequency dividers.** Because J-K flip-flops that are in the toggle mode (J = 1, K = 1) change state only on each negative (or positive) edge of the clock pulse, the **frequency** of a flip-flop output waveform is one-half the frequency of the waveform at its clock input. Therefore, each flip-flop in an asynchronous (ripple) counter will divide the frequency output of the preceding flip-flop by two. Because the number of counter binary states (modulus) is multiplied by two for each additional flip-flop, the **divide-by of the last counter flip-flop output** is always equal to the **modulus (MOD)** of the counter.

The circuit in Figure 25-1 consists of four negative-edge-triggered J-K flip-flops wired as a **4-bit asynchronous (ripple) binary counter**. Notice that the flip-flop J and K inputs are connected to 5 V (J = 1, K = 1), causing the flip-flops to **toggle** when they are clocked with a negative clock edge. The first flip-flop (Q0) CLK input is used as the counter CLK' input. Each successive flip-flop CLK input is connected to the preceding flip-flop output (Q) (right-to-left). When the preceding flip-flop output (Q) drops from binary one (1) to binary zero (0), it will toggle the next flip-flop. The counter **active low CLR' input** is formed by connecting all of the flip-flop active low dc clear inputs together. When the CLR' input is dropped to binary zero (0), all of the flip-flops will clear (0), causing the counter to be cleared. The counter CLR' input must be returned to binary one (1) in order for the counter to count. If the active low CLR' input is not returned to binary one (1), the counter will stay cleared because the flip-flop dc clear inputs **override** the flip-flop CLK inputs.

The circuit in Figure 25-2 will demonstrate the effect of flip-flop **propagation delay time (t_p)** on the timing of a 4-bit asynchronous (ripple) counter. The word generator data ready (R) will apply the clock pulses to the counter CLK' input. The word generator will also clear the counter by applying a zero (0) to the CLR' input on the first clock pulse, and then it will raise the CLR' input to one (1) for the rest of the count. The logic analyzer will monitor the counter clock input (CLK') and the counter outputs (Q3–Q0).

The circuit in Figure 25-3 consists of four negative-edge-triggered J-K flip-flops wired as a **4-bit asynchronous (ripple) binary down counter**. Notice that each successive flip-flop CLK input (right-to-left) is connected to the preceding flip-flop **inverted output**. This causes the counter to **count down** instead of counting up. The remainder of the circuit is the same as the up counter in Figure 25-1.

The 74293 (7493) in Figure 25-4 is an **asynchronous (ripple) binary counter**. It has four J-K flip-flops wired in the toggle mode, with flip-flop outputs QA, QB, QC, and QD. It is wired internally as a **1-bit (MOD-2 or divide-by-2) counter** with a negative-edge triggered clock input INA and output QA, and a **3-bit (MOD-8 or divide-by-8) counter** with negative-edge triggered clock input INB and outputs QB, QC, and QD. The 74293 (7493) can be externally wired as a **4-bit (MOD-16 or divide-by-16) counter** by connecting the output of the 1-bit counter (QA) to the clock input of the 3-bit counter (INB). This feature makes the 74293 (7493) very versatile because it can be wired as a **1-bit, 3-bit, or 4-bit counter**. The 74293 (7493) also has two active high reset (clear) inputs (RO1 and RO2). These inputs can be used to reset (clear) the counter or to cut off the binary count in order to change the modulus (divide-by) of the counter. By wiring the counter in different configurations, the modulus (divide-by) of the counter can be changed to any value between MOD-2 (divide-by-2) and MOD-16 (divide-by-16).

The circuit in Figure 25-5 will display the **timing** of the 74293 (7493) **asynchronous (ripple) counter**. The word generator data ready output will apply the clock pulses to the counter clock inputs. The logic analyzer will monitor the clock pulses and the counter outputs (QA–QD). When the D switch is down, the 74293 (7493) is configured as a **1-bit (MOD-2 or divide-by 2) counter** using the INA clock input and a **3-bit (MOD-8 or divide-by-8) counter** using the INB clock input. When switch D is up, the 74293 (7493) is configured as a **4-bit (MOD-16 or divide-by-16) counter** using INA as the clock input with output QA connected to INB. The active high reset (clear) inputs (RO1 and RO2) are permanently connected to ground (0) to allow the counter to count.

The 74293 (7493) in Figure 25-6 is wired as a **MOD-12 (divide-by-12) counter**. Notice that it is wired as a 4-bit counter (QA is connected to INB) and outputs QD and QC are connected to reset (clear) inputs RO1 and RO2. This will cause the counter to count between zero and eleven (12 states) and reset (clear) on the twelfth clock pulse (binary output 1100), making it a MOD-12 (divide-by-12) counter.

The 74293 (7493) counters in Figure 25-7, 25-8, and 25-9 are connected as **divide-by counters**. The clock input and the counter output will be displayed on the oscilloscope screen. You can determine the expected **modulus (divide-by)** of these counters by observing the wiring of QA, INB, RO1, and RO2 and following the reasoning used in the discussion for Figure 25-6. You can measure the modulus (divide-by) of these counters by measuring the **time period (T)** for one cycle of the clock pulse (T_c) and the **time period (T)** for one cycle of the counter output (T_o). From these values, you can calculate the clock frequency (f_c) and the counter output frequency (f_o) using the equation

$$f = \frac{1}{T}$$

The modulus (divide-by) is calculated from

$$\text{Modulus} = \frac{f_c}{f_o}$$

The circuit in Figure 25-10 consists of two 74293 (7493) **cascaded asynchronous counters** to increase the modulus (divide-by). The function generator will apply a 100 kHz pulse frequency to the clock input. The oscilloscope will monitor the cascaded counter output. You can determine the expected modulus (divide-by) of the cascaded counters by first determining the expected modulus (divide-by) of each of the two counters. The MOD number (divide-by) of the cascaded pair is equal to the product of their individual MOD numbers. The output frequency (f_o) is then determined by dividing the input clock frequency by the modulus (divide-by) of the cascaded pair.

Figure 25-1 4-Bit Asynchronous (Ripple) Counter

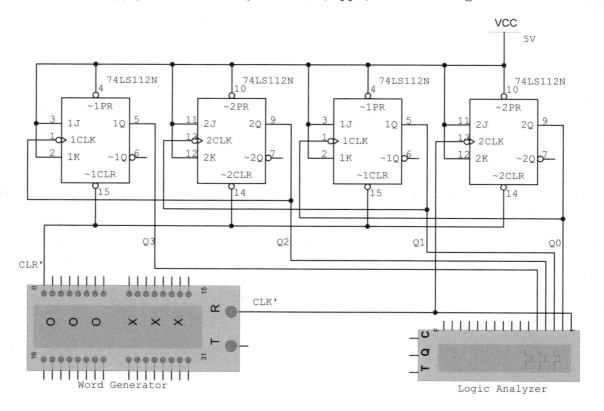

Figure 25-2 4-Bit Asynchronous (Ripple) Counter Timing

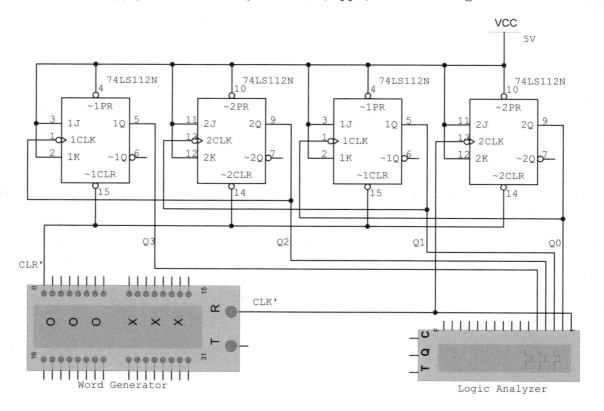

Figure 25-3 4-Bit Asynchronous (Ripple) Down Counter

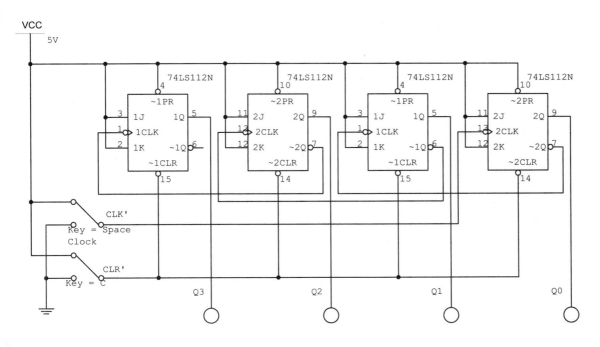

Figure 25-4 74293 Asynchronous (Ripple) Counter

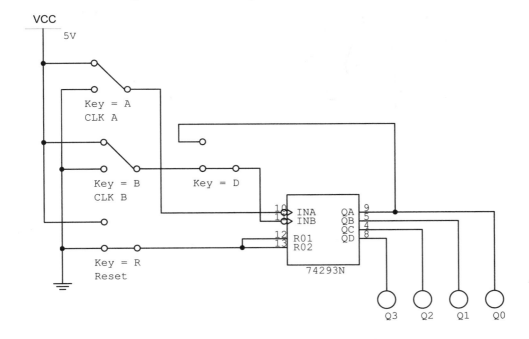

Figure 25-5 74293 Asynchronous (Ripple) Counter Timing

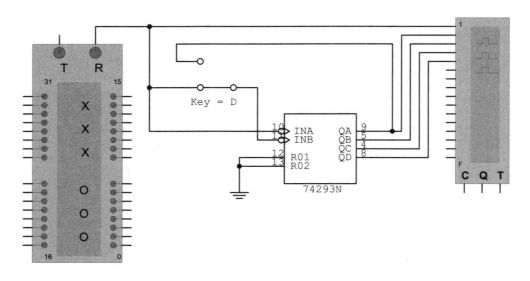

Figure 25-6 74293 Wired as a MOD-12 Counter

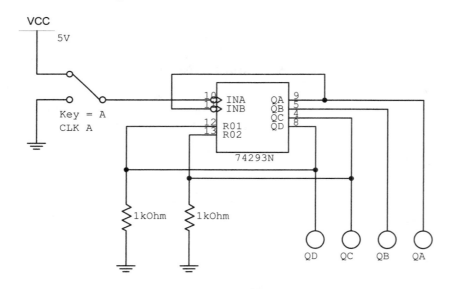

Figure 25-7 74293 Frequency Division

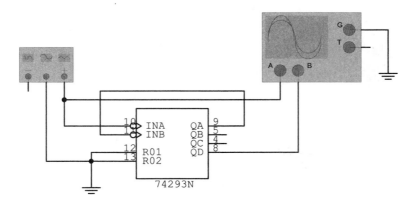

Figure 25-8 74293 Frequency Division

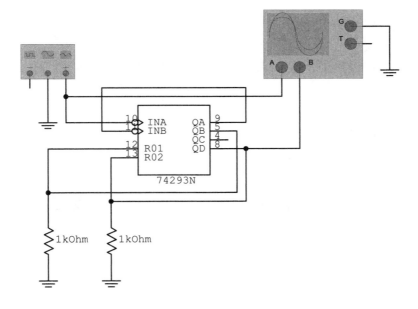

Figure 25-9 74293 Frequency Division

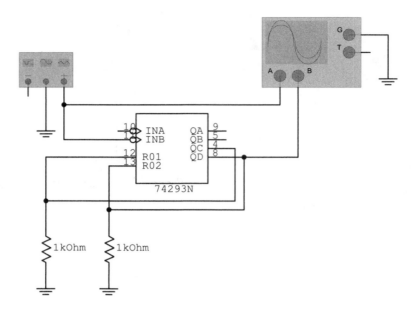

Figure 25-10 74293 Frequency Division

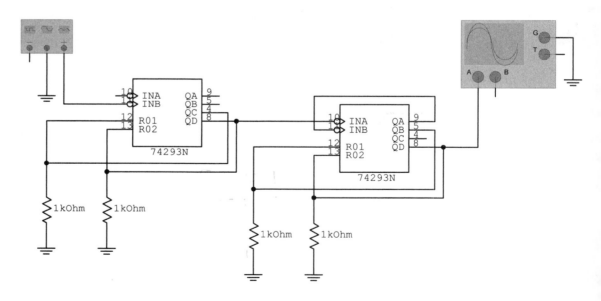

Procedure:

Step 1.   Open circuit file FIG25-1. You are looking at four negative-edge-triggered J-K flip-flops wired as a 4-bit asynchronous (ripple) binary counter. The C and CLK switches should be up (1). Click the On-Off switch to run the simulation and press the C key on the computer keyboard to clear the counter. Press the C key again to raise the CLR′ back to binary one (1).

Questions: Is the clear input (CLR′) active low or active high?

ACTIVE   LOW

Why does the CLR′ input need to be binary one (1) in order for the counter to be able to count?

BECAUSE IT'S AN ACTIVE LOW, AND WHEN IT'S "O", IT IS CLEARING IT. No changes.

Step 2.   Press the space bar on the computer keyboard to make the CLK′ input drop from one (1) to zero (0) to produce a negative clock edge. Record the counter binary output for one clock pulse in Table 25-1.

Table 25-1 Output Count

Clock Pulse	Output Q3 Q2 Q1 Q0
1	0 0 0 1
2	0 0 1 0
3	0 0 1 1
4	0 1 0 0
5	0 1 0 1
6	0 1 1 0
7	0 1 1 0
8	1 0 0 0
9	1 0 0 1
10	1 0 1 0
11	1 0 1 1
12	1 1 0 0
13	1 1 0 1
14	1 1 1 0
15	1 1 1 1
16	0 0 0 0

Question: Is the counter positive-edge-triggered or negative-edge-triggered?

NEGATIVE

Step 3.   Press the space bar on the computer keyboard enough times to produce another negative clock edge. Record the counter binary output for two clock pulses in Table 25-1. Repeat this procedure until the table is complete, then stop the simulation.

Questions: Based on the data in Table 25-1, what conclusions can you draw about the relationship between the counter binary output and the number of clock pulses applied to the counter CLK' input?

IT'S A COUNTER WITH NEGATIVE EDGE, IT NEEDS TWO CLK PULSES TO GET 1 COUNT.

What happened when the counter was clocked after the maximum count was reached?

IT STARTED OVER AGAN, REACHING 0000 AND STARTING FROM 0001 AGAIN

Step 4.   Open circuit file FIG25-2. You are looking at a circuit that will demonstrate the effect of flip-flop propagation delay time (t_p) on the timing of a 4-bit asynchronous (ripple) counter. Bring down the word generator enlargement and make sure that the following settings are selected: Frequency = 1 kHz, Trigger = Internal, Controls = Burst, Hex codes (00000000 once, then 00000001 repeated fifteen times). Move the word generator to the left. Bring down the logic analyzer enlargement and make sure that the following settings are selected: Clocks/Div = 16, Clock Setup (Clock Source = Internal, Clock Rate = 10 kHz, Pre-trigger Samples = 100, Post-trigger Samples = 1000, Threshold Voltage = 2.5), Trigger Settings (Trigger Clock Edge = Positive, Trigger Qualifier = x, Pattern A= xxxxxxxxxxxxxxxx, Trigger Combinations = A). Move the logic analyzer to the right. Click the On-Off switch to run the simulation. The light blue curve plot on the logic analyzer screen is the clock (CLK') input to the counter. The dark blue (Q0), dark green (Q1), brown (Q2), and light green (Q3) curve plots represent the 4-bit counter output (Q3–Q0). Draw and label the CLK' input and the counter output waveshapes in the space provided, then stop the simulation by clicking the On-Off switch.

NOTE: If this experiment is being performed in a hardwired laboratory, use two 74112 ICs for the four negative-edge-triggered J-K flip-flops and use a pulse generator in place of the word generator. If a logic analyzer is not available, use a dual-trace oscilloscope.

Questions: Is each output changing state on the positive or the negative edge of the previous flip-flop output pulse? **Explain why.**

ON THE NEGATIVE.

BECAUSE IT'S A NEGATIVE EDGE TRIGGERED CKT.

What is the relationship between the frequencies of the flip-flop output pulses?

2:1 FROM EACH OTHER, FROM Q_0 TO Q_3

What is the relationship between the frequency of each flip-flop output pulse and the CLK' input pulse?

$Q_0 = 2:1$
$Q_1 = 4:1$
$Q_2 = 8:1$
$Q_3 = 16:1$

Are the output binary counts correct after each clock input pulse (negative edge)?

DISCARDING THE 1st, YES

Step 5.   Change the frequency of the word generator to 20 MHz. Reset the cursor on the word
              generator to the top of the column of numbers by right clicking the top number and selecting
              "Set Cursor." Change the internal clock rate on the logic analyzer to 100 MHz and change
              the Clocks/Div to 8. Click the On-Off switch to run the simulation. Draw and label the new
              CLK′ input and counter output waveshapes in the space provided, then stop the simulation.

Questions: Are all of the counter output pulses changing state on the negative edge of the previous
flip-flop pulse? **Explain.**

No. THEY ARE MESSED UP BECAUSE OF THE PROPAGATION DELAY
ON THE 4 FLIP FLOPS.

Are the output binary counts still correct after all of the clock inputs (negative edges)? **Explain.**

No, THEY'RE MESSED UP.

Step 6.   Use the cursors on the logic analyzer to measure the propagation delay time (t_p) between
              the negative edge of a pulse and the next FF pulse positive edge and record your answer.

$t_p = \underline{14.4 \text{ NS}}$

Step 7.   Calculate the counter maximum clock frequency (f_{max}) based on the flip-flop propagation
              delay time (t_p) and the number of counter flip-flops (N).

$$f_{max} = \frac{1}{(N \cdot T_P)} \Longrightarrow 17.36 \text{ MHz}$$

Question: Did the clock frequency of the word generator exceed the maximum clock frequency (f_{max})?

Yes, 20 MHz

Step 8.   Open circuit file FIG 25-3. You are looking at four negative-edge-triggered J-K flip-flops wired as a 4-bit asynchronous (ripple) binary **down** counter. The C and CLK switches should be up (1). Click the On-Off switch to run the simulation and press the C key on the computer keyboard to clear the counter. Press the C key again to raise CLR' back to binary one (1).

Step 9.   Press the space bar on the computer keyboard to make the CLK' input drop from one (1) to zero (0) to produce a negative clock edge. Record the counter binary output for one clock pulse in Table 25-2. Press the space bar on the computer keyboard enough times to produce another negative clock edge. Record the counter binary output for two clock pulses in Table 25-2. Repeat this procedure until the table is complete, then stop the simulation.

Table 25-2 Output Count

Clock Pulse	Output Q3 Q2 Q1 Q0
1	
2	
3	
4	
5	
6	
7	
8	
9	
10	
11	
12	
13	
14	
15	
16	

Question: Based on the data in Table 25-2, what conclusion can you draw about the direction of the binary count as the clock pulses are applied to the counter CLK′ input?

Step 10.   Open circuit file FIG 25-4. You will test a 74293 (7493) asynchronous (ripple) counter. This counter has two active high reset (clear) inputs (RO1 and RO2) and two negative-edge-triggered clock inputs (INA and INB). The INA input is used to trigger flip-flop A and the INB input is used to trigger the 3-bit counter using flip-flops B, C, and D. The 74293 (7493) can be externally wired as a 4-bit (MOD-16) counter by connecting flip-flop output QA to clock input INB and using INA as the clock input. When the D switch is down, the 74293 (7493) is configured as a 1-bit (MOD-2) counter using the INA input and a 3-bit (MOD-8) counter using the INB input. When the D switch is up, the 74293 (7493) is configured as a 4-bit (MOD-16) counter using the INA input. See the Theory section for more details about the 74293 (7493) asynchronous (ripple) counter.

Step 11.   Switches R and D should be down (0) and the CLKA and CLKB switches should be up (1). Click the On-Off switch to run the simulation. If the counter output (Q3–Q0) is not cleared, press the R key on the computer keyboard to raise the reset (clear) input (RO1 and RO2) to binary one (1) to clear the counter. Press the R key again to lower the reset input back to binary zero (0).

Step 12.   With switch D down, press the A key on the computer keyboard to produce a negative clock edge on the INA input. Keep pressing the A key to produce a number of negative clock edges on the INA input.

Questions: What did you observe at counter output Q0? **Explain.**

Step 13.   With switch D down, press the B key on the computer keyboard to produce a negative clock edge on the INB input. Keep pressing the B key to produce a number of negative clock edges on the INB input.

Questions: What did you observe at the counter outputs (Q3–Q1)? **Explain.**

Step 14.   Press the D key on the computer keyboard to raise switch D up. This will connect output
            QA to input INB. Press the R key to reset (clear) the counter, and then press it again to
            lower it to zero (0). Keep pressing the A key to apply a number of negative clock edges to
            the INA input and observe the output (Q3–Q0), then stop the simulation.

Question: What did you observe at the counter outputs (Q3–Q0)? **Explain.**

Step 15.   Open circuit file FIG25-5. You are looking at a circuit that will display the timing of a
            74293 (7493) asynchronous (ripple) counter. The word generator and logic analyzer
            settings should be as shown in Step 4. When the D switch is down, the 74293 (7493) is
            configured as a 1-bit (MOD-2) counter using the INA input and a 3-bit (MOD-8) counter
            using the INB input. When the D switch is up, the 74293 (7493) is configured as a 4-bit
            (MOD-16) counter using the INA input. Switch D should be down.

> NOTE: If this experiment is being performed in a hardwired laboratory, use a pulse generator in
> place of the word generator. If a logic analyzer is not available, use a dual-trace oscilloscope.

Step 16.   Click the On-Off switch to run the simulation. The light blue curve plot on the logic
            analyzer screen is the clock input (INA and INB). The dark blue curve plot represents the
            1-bit (MOD-2) counter output (QA) clocked on the INA input. The dark green (QB), brown
            (QC), and light green (QD) curve plots represent the 3-bit (MOD-8) counter output clocked
            on the INB input. Draw and label the curve plots in the space provided, then stop the
            simulation.

Questions: What is the frequency relationship between the INA input (light blue) and the counter QA output (dark blue)? What is the modulus (divide-by) of this counter?

What is the frequency relationship between the INB input (light blue) and the counter outputs QB (dark green), QC (brown), and QD (light green)? What is the modulus (divide-by) of this counter?

Step 17.   Press the D key on the computer keyboard to raise switch D. This will connect output QA to input INB. Reset the cursor on the word generator to the top of the column of numbers (see Step 5). Click the On-Off switch to run the simulation again. Draw and label the new curve plots in the space provided, then stop the simulation.

Question: What is the frequency relationship between the INA input (light blue) and counter outputs QA (dark blue), QB (dark green), QC (brown), and QD (light green)? What is the modulus (divide-by) of this counter?

Step 18. Open circuit file FIG 25-6. You will test a 74293 (7493) wired as a MOD-12 (divide-by-12) counter. Notice that output QA is connected to the INB input, output QD is connected to reset input RO1, and output QC is connected to reset input RO2. Click the On-Off switch to run the simulation. Press the A key on the computer keyboard enough times to apply a series of negative clock edges to the INA input and observe the output (QD–QA). After the output reaches zero (0000), continue pressing the A key and record the results in Table 25-3, then stop the simulation.

Table 25-3 Output Count

Clock Pulse	Output QD QC QB QA
1	
2	
3	
4	
5	
6	
7	
8	
9	
10	
11	
12	
13	
14	
15	
16	

Questions: What conclusion can you draw about the relationship between the counter binary output and the clock pulses applied to the INA input?

What happened on count number twelve? **Explain**.

Based on these results, what is the modulus (divide-by) of this counter? Does it match the expected value based on the circuit configuration?

Step 19.   Open circuit file FIG25-7. Bring down the function generator enlargement and make sure that the following settings are selected: *Square Wave*, Freq = 10 kHz, Duty Cycle = 50%, Ampl = 2.5 V, Offset = 2.5 V. Move the function generator to the left. Bring down the oscilloscope enlargement and make sure that the following settings are selected: Time base (Scale = 500 μs/Div, Xpos = 0, Y/T), Ch A (Scale = 5 V/Div, Ypos = 0, DC), Ch B (Scale = 5 V/Div, Ypos = –2, DC), Trigger (Pos. edge, Level = 1 μV, Sing, A). Move the oscilloscope to the right. Click the On-Off switch to run the simulation. After one screen display, stop the simulation. Use the cursors on the oscilloscope to determine the time period (T) for one cycle and the frequency (f) of the clock input (red) and the counter output (blue). Record your answers in the space provided.

Clock input $T_c =$ _____

                       $f_c =$ _____

Counter output $T_o =$ _____

                       $f_o =$ _____

Step 20.   Based on the results in Step 19, calculate the modulus (divide-by) of this counter.

Questions: Based on the circuit configuration in Figure 25-7, what is the expected modulus (divide-by) of this counter? **Explain why.**

How does this answer compare with the calculated modulus in Step 20?

Step 21.   Open circuit file FIG 25-8. The function generator and oscilloscope settings should be the same as in Step 19, except the oscilloscope Time base scale should be 200 μs/Div. Click the On-Off switch to run the simulation. After one full screen display, stop the simulation. Use the cursors on the oscilloscope to determine the time period (T) for one cycle and the frequency (f) of the clock input (red) and the counter output (blue). Record your answers in the space provided.

Clock input $T_c =$ _____

                       $f_c =$ _____

Counter output $T_o =$ _____

                       $f_o =$ _____

Step 22.   Based on the results in Step 21, calculate the modulus (divide-by) of this counter.

Questions: Based on the circuit configuration in Figure 25-8, what is the expected modulus (divide-by) of this counter? **Explain.**

How does this answer compare with the calculated modulus in Step 22?

Step 23.   Open circuit file FIG 25-9. The function generator and oscilloscope settings should be the
            same as in Step 21. Click the On-Off switch to run the simulation. After one full screen
            display, stop the simulation. Use the cursors on the oscilloscope to determine the time
            period (T) for one cycle and the frequency (f) of the clock input (red) and the counter
            output (blue). Record your answer in the space provided.

                        Clock input $T_c =$ _____

                                                     $f_c =$ _____

                        Counter output $T_o =$ _____

                                                     $f_o =$ _____

Step 24.   Based on the results in Step 23, calculate the modulus (divide-by) of this counter.

Questions: Based on the circuit configuration in Figure 25-9, what is the expected modulus (divide-by) of this counter? **Explain.**

How does this answer compare with the calculated modulus in Step 24?

Step 25. Open circuit file FIG 25-10. Bring down the function generator enlargement and make sure that the following settings are selected: *Square Wave*, Freq = 100 kHz, Duty Cycle = 50%, Ampl = 2.5 V, Offset = 2.5 V. Move the function generator to the left. Bring down the oscilloscope enlargement and make sure that the following settings are selected: Time base (Scale = 200 μs/Div, Xpos = 0, Y/T), Ch A (Scale = 5 V/Div, Ypos = 0, DC), Ch B (Scale = 5 V/Div, Ypos = 0, DC), Trigger (Pos. edge, Level = 1 μV, Sing, A). Move the oscilloscope to the right. Click the On-Off switch to run the simulation. After a full screen display, stop the simulation. Use the cursors on the oscilloscope to determine the time period (T_o) for one cycle and the frequency (f_o) of the counter output (blue). Record your answer in the space provided.

$T_o =$ _____

$f_o =$ _____

Step 26. Based on the cascaded counter circuit in Figure 25-10 and the 100 kHz clock frequency (f_c), determine the expected output frequency (f_o).

Question: How did your calculated output frequency (f_o) compare with the value measured in Step 25?

EXPERIMENT

26 Synchronous Counters

Objectives:

1. Demonstrate how J-K flip-flops are used to build a synchronous binary counter.
2. Demonstrate the effect of propagation delay on synchronous counter timing.
3. Demonstrate synchronous up/down counting.
4. Investigate the operation of a 74191 presettable up/down synchronous binary counter.
5. Demonstrate frequency division using synchronous counters.
6. Learn how to change the modulus (divide-by) of a synchronous counter without changing the wiring.
7. Demonstrate cascaded synchronous counters.

Materials:

One 5 V dc power supply
Eight logic switches
Ten logic probe lights
Four negative-edge-triggered J-K flip-flops (2-74112 ICs)
Two presettable up/down synchronous binary counters (2-74191 ICs)
Four two-input AND gates (1-7408 IC)
Two two-input OR gates (1-7432 IC)
One two-input NAND gate (1-7400 IC)
One INVERTER (1-7404 IC)
One function (pulse) generator
One logic analyzer or dual-trace oscilloscope

Theory:

Make sure you complete Experiment 25 before attempting this experiment. Also, review the Theory section of Experiment 25.

Synchronous counters eliminate the problem of accumulated propagation delay encountered in asynchronous counters because all of the flip-flops receive the input clock pulses **simultaneously** in synchronous counters. Therefore, some means must be used to control when a flip-flop will toggle and when it will not toggle in a synchronous counter. This is accomplished with **logic networks** connected between the J-K flip-flop inputs and the previous flip-flop outputs. Due to the additional logic circuitry, synchronous counters have **more complicated circuitry** and **higher cost** than asynchronous counters.

Because of the elimination of the accumulated **propagation delay time (t_p)** in synchronous counters, the time delay between the counter clock input and the response of the last flip-flop output does not depend on the number of counter flip-flops. For this reason, the **maximum clock frequency (f_{max})** for synchronous counters is **higher** than the maximum clock frequency for asynchronous counters. For a synchronous counter, the maximum clock frequency can be calculated from

$$f_{max} \cong \frac{1}{t_p}$$

neglecting the propagation delay of the AND gates.

Synchronous counters are also often used as frequency dividers. Each output in a synchronous counter is equal to one-half the frequency of the preceding output and the number of synchronous **counter binary states (modulus)** is multiplied by two for each additional output. Therefore, the **divide-by of the last counter output** is equal to the **modulus (MOD)** of the counter.

The circuit in Figure 26-1 consists of four **negative-edge-triggered J-K flip-flops** wired as a **4-bit synchronous binary counter**. Notice that all of the flip-flop clock inputs are connected to the counter CLK' input. Each flip-flop J and K input is connected to either 5 V (J = 1, K = 1), the output of another flip-flop, or the output of a logic gate (AND gate). The correct logic causes the flip-flops to toggle at the proper time to produce a binary up count. The counter **active low CLR' input** is formed by connecting all of the flip-flop active low CLR inputs together. When the counter CLR' input is dropped to binary zero (0), all of the flip-flops will clear (0), causing the counter to be cleared. The counter CLR' input must be returned to binary one (1) in order for the counter to count. If the active low CLR' input is not returned to binary one (1), the counter will stay cleared because the flip-flop CLR inputs override the flip-flop CLK inputs.

The circuit in Figure 26-2 will demonstrate the effect of **flip-flop propagation delay time (t_p)** on the timing of a 4-bit synchronous binary counter. The word generator will apply the clock pulses to the counter CLK' input. The word generator will also clear the counter by applying a zero (0) to the counter CLR' input on the first clock pulse, and then it will raise the counter CLR' input to one (1) for the rest of the count. The logic analyzer will monitor the clock input (CLK') and the counter outputs (Q3–Q0).

The circuit in Figure 26-3 consists of three **negative-edge-triggered J-K flip-flops** wired as a **3-bit synchronous up/down binary counter**. Notice that the **down count logic circuitry** (bottom AND gates) is identical to the **up count logic circuitry** (top AND gates), except the down count logic is connected to the inverted flip-flop outputs. When the UP/DOWN' switch is up, the up count logic gates are opened and the down count logic gates are closed. When the UP/DOWN' switch is down, the up count logic gates are closed and the down count logic gates are opened.

The 74191 in Figure 26-4 is a **presettable up/down synchronous binary counter**. It can be preset to any count by applying a 4-bit binary number to inputs DCBA and dropping the **active low load input (LOAD)** to binary zero (0). The load input (LOAD) must be returned to binary one (1) in order for the counter to count because the load input **overrides** the CLK input. The 74191 will count up if a binary zero (0) is applied to the **U'/D input** and it will count down if a binary one (1) is applied to the U'/D input. The counter will increment or decrement one count each time the CLK input receives a positive clock edge. The active low **count enable input (CTEN)** must be low (0) for the counter to be enabled, otherwise it will be disabled. The **terminal count (MAX/MIN) output** is normally low (0) but will go high (1) for one clock period when the counter reaches zero (0000) in the count down mode or fifteen

(1111) in the count up mode. The **ripple clock (RCO) output** is normally high (1), but will go low (0) for the low part of the clock period when the counter reaches zero (0000) in the count down mode or fifteen (1111) in the count up mode. The MAX/MIN and RCO outputs are made available in order to be able to **cascade** 74191 synchronous counters to provide a count higher than fifteen (1111). Each additional cascaded counter adds 4 bits to the output.

The circuit in Figure 26-5 will display the **timing** of the **74191 synchronous binary counter**. The word generator will apply the clock pulses to the counter clock input. It will also apply a binary zero (0) to the load input (LOAD) on the first clock pulse and a binary one (1) for the rest of the count. This will cause the counter to start the count at the binary number set by the DCBA switches. The logic analyzer will monitor the clock pulses, the load input (LOAD), the counter outputs (QA– QD), the ripple clock output (RCO), and the terminal count output (MAX/MIN).

The 74191 in Figure 26-6 is wired as a **MOD-12 (divide-by-12) counter**. Notice that the MAX/MIN output and the CLK input are connected to a NAND gate that feeds the LOAD input. This will cause the counter to **reset** the count during the **positive part of the clock period** when the counter output reaches the terminal state and the **MAX/MIN output is high (1)**. The counter will **reset** to the binary count that is set by the DCBA switches. (Up equals a binary one and down equals a binary zero on the DCBA switches.) Because this counter is in the **count down mode**, the binary number represented by the **DCBA switches** will determine the **modulus** (number of binary states) of the counter. Therefore, the modulus (MOD) of this counter can be changed by simply changing the DCBA switches, eliminating the need to change the wiring.

The **74191 synchronous counter** in Figure 26-7, is connected as a **divide-by counter**. The clock input and the counter output will be displayed on the oscilloscope screen. You can determine the expected **modulus (divide-by)** of this counter by observing the position of the DCBA switches (up equals binary one and down equals binary zero) and following the reasoning used in the discussion for Figure 26-6. You can measure the modulus (divide-by) of this counter by measuring the **time period (T)** for one cycle of the **clock pulse (T_c)** and the **time period (T)** for one cycle of the **counter output (T_o)**. From these values, you can calculate the **clock frequency (f_c)** and the **counter output frequency (f_o)** using the equation

$$f = \frac{1}{T}$$

The modulus (divide-by) is calculated from

$$\text{Modulus} = \frac{f_c}{f_o}$$

The circuit in Figure 26-8 consists of two **74191 cascaded synchronous counters** wired as an **8-bit counter**. The counter on the right outputs the low 4-bits and the counter on the left outputs the high 4-bits. The cascaded 8-bit synchronous counter can **count up or down** depending on the position of the UP'/DOWN switch. The RCO output of the counter on the right is connected to the active low enable input (CTEN) of the counter on the left. This will cause the counter on the left to be enabled during the low part of the clock period and increment or decrement one count when the counter on the right reaches its terminal count and begins a new count cycle. Notice that both counters are being clocked at the same time, making the cascaded pair **synchronous**.

Figure 26-1 4-Bit Synchronous Counter

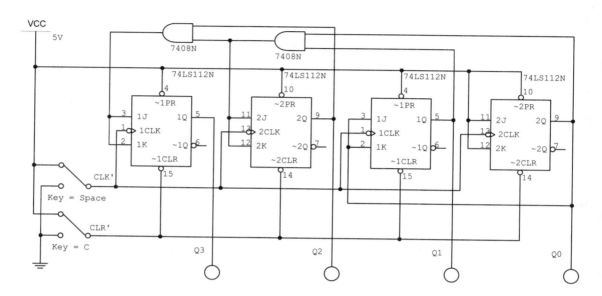

Figure 26-2 4-Bit Synchronous Counter Timing

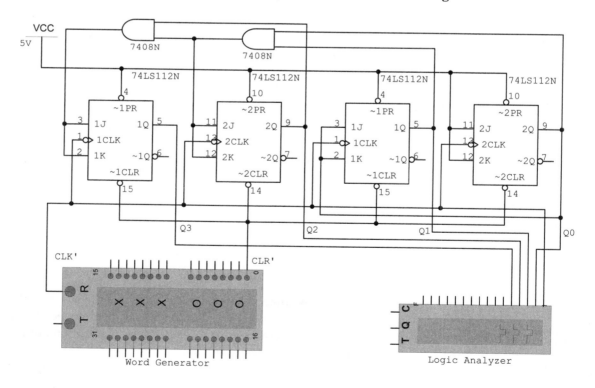

Figure 26-3 3-Bit Synchronous Up/Down Counter

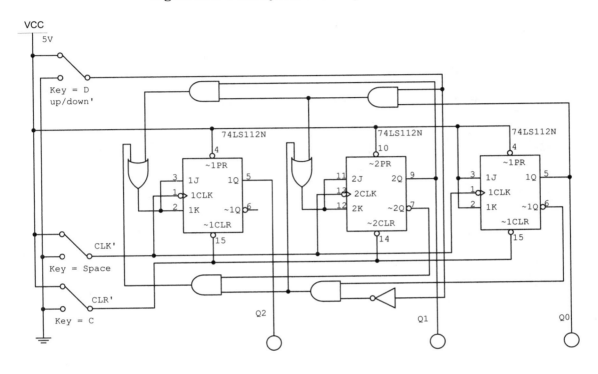

Figure 26-4 74191 Presettable Up/Down Synchronous Binary Counter

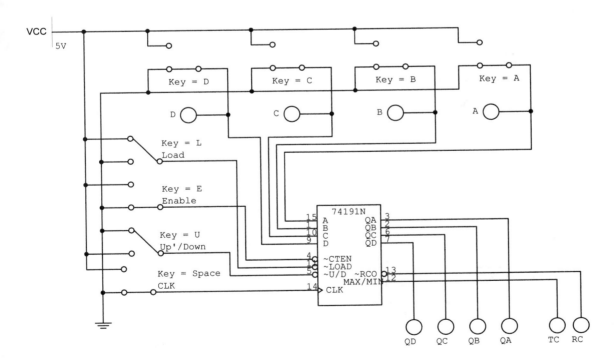

Figure 26-5 74191 Synchronous Counter Timing

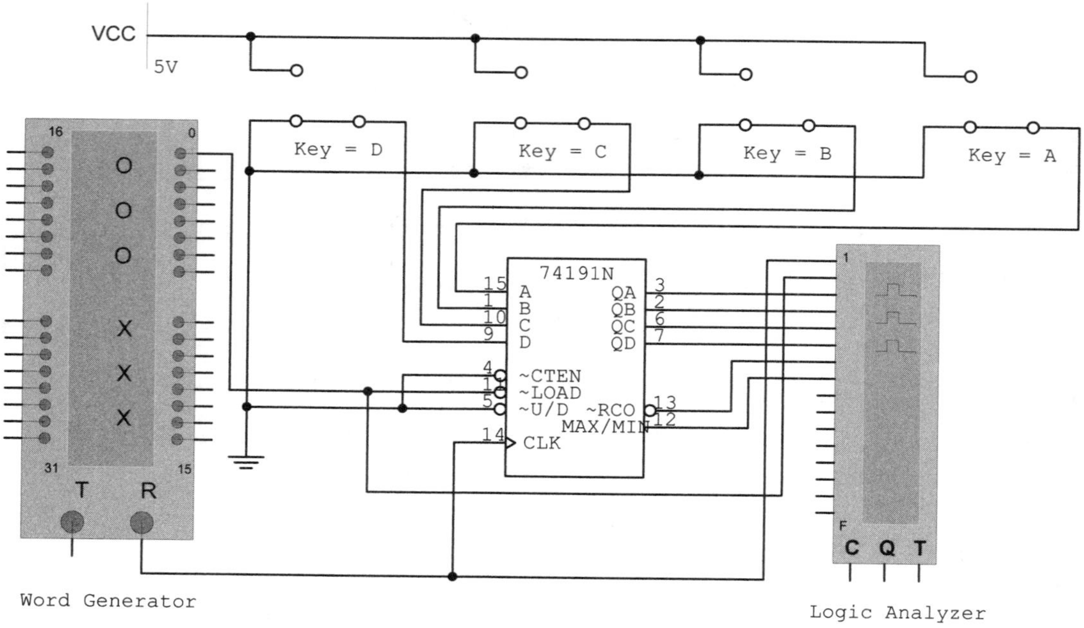

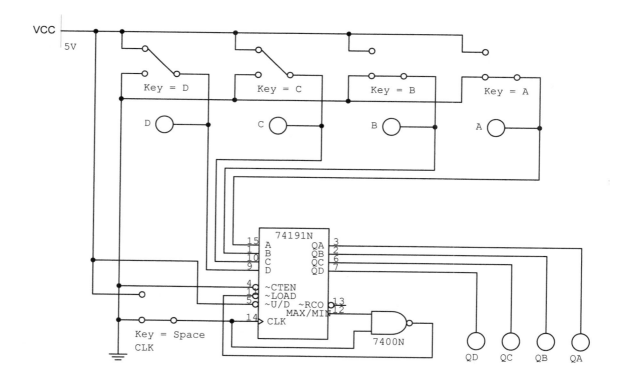

Figure 26-6 74191 Wired as a MOD-12 Counter

Figure 26-7 74191 Frequency Division

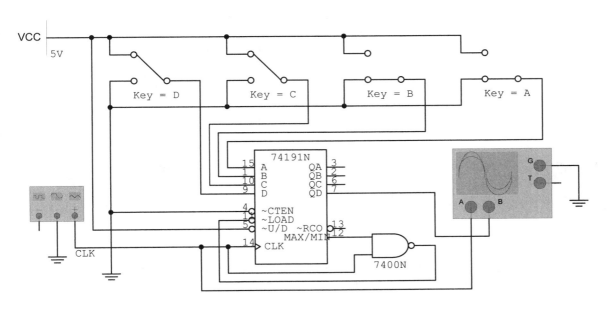

Figure 26-8 Cascaded 74191 Synchronous Counters (8-Bit Counter)

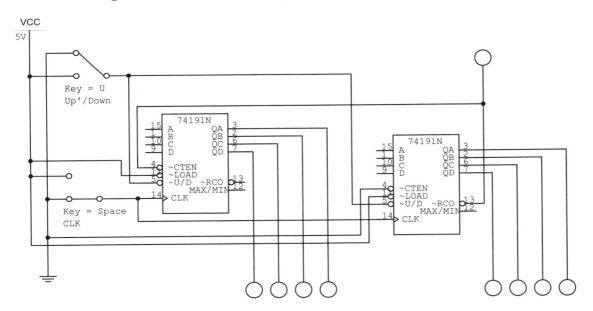

Procedure:

Step 1.   Open circuit file FIG26-1. You are looking at four negative-edge-triggered J-K flip-flops
           wired as a 4-bit synchronous binary counter. The C and CLK switches should be up (1).
           Click the On-Off switch to run the simulation and press the C key on the computer keyboard
           to clear the counter. Press the C key again to raise the CLR' back to binary one (1).

Questions: Is the clear input (CLR') active low or active high?

Why does the CLR' input need to be binary one (1) in order for the counter to be able to count?

Step 2.   Press the space bar on the computer keyboard to make the CLK' input drop from one (1) to
           zero (0) to produce a negative clock edge. Record the counter binary output for one clock
           pulse in Table 26-1.

Table 26-1 Output Count

Clock Pulse	Output Q3 Q2 Q1 Q0
1	
2	
3	
4	
5	
6	
7	
8	
9	
10	
11	
12	
13	
14	
15	
16	

Question: Is the counter positive-edge-triggered or negative-edge-triggered?

Step 3.   Press the space bar on the computer keyboard enough times to produce another negative clock edge. Record the counter binary output for two clock pulses in Table 26-1. Repeat this procedure until the table is complete, then stop the simulation.

Questions: Based on the data in Table 26-1, what conclusion can you draw about the relationship between the counter binary output and the number of clock pulses applied to the counter CLK′ input?

What happened when the counter was clocked after the maximum count was reached?

Step 4.   Pull down the File menu and open FIG26-2. You are looking at a circuit that will demonstrate the effect of flip-flop propagation delay time (t_p) on the timing of a 4-bit synchronous binary counter. Bring down the word generator enlargement and make sure that the following settings are selected: Frequency = 1 kHz, Trigger = Internal, Controls = Burst, Hex codes (00000000 once, then 00000001 repeated fifteen times). Move the word generator to the left. Bring down the logic analyzer enlargement and make sure that the following settings are selected: Clocks/Div = 16, Clock Setup (Clock Source = Internal, Clock Rate = 10 kHz, Pre-trigger Samples = 100, Post-trigger Samples = 1000, Threshold Voltage = 2.5), Trigger Settings (Trigger Clock Edge = Positive, Trigger Qualifier = x, Pattern A= xxxxxxxxxxxxxxxx, Trigger Combinations = A). Move the logic analyzer to the right. Click the On-Off switch to run the simulation. The light blue curve plot on the logic analyzer screen is the clock (CLK′) input to the counter. The dark blue (Q0), dark green (Q1), brown (Q2), and light green (Q3) curve plots represent the 4-bit counter output (Q3–Q0). Draw and label the CLK′ input and the counter output waveshapes in the space provided, then stop the simulation.

NOTE: If this experiment is being performed in a hardwired laboratory, use two 74112 ICs for the four negative-edge-triggered J-K flip-flops and one 7408 IC for the 2-input AND gates. Use a pulse generator in place of the word generator. If a logic analyzer is not available, use a dual-trace oscilloscope.

Questions: What is the relationship between the frequencies of the flip-flop output pulses?

What is the relationship between the frequency of each flip-flop output pulse and the CLK′ input pulse?

Are the output binary counts correct after each clock input pulse (negative edge)?

Step 5.   Change the frequency of the word generator to 20 MHz and reset the cursor to the top of the column of numbers by right clicking the top number and selecting "Set Cursor." Change the internal clock rate on the logic analyzer to 100 MHz and change the Clocks/Div to 8. Click the On-Off switch to run the simulation. Draw and label the new CLK′ input and counter output waveshapes in the space provided, then stop the simulation.

Question: Are the output binary counts correct after each of the clock inputs (negative edges)? **Explain**.

Step 6.   Calculate the counter maximum clock frequency (f_{max}) based on the flip-flop propagation delay time. (See Experiment 25, Step 6, for the value of the flip-flop delay time.)

Question: How does f_{max} for the synchronous counter compare with f_{max} for the asynchronous counter in Experiment 25, Step 7?

How does f_{max} for the synchronous counter compare with the frequency on the word generator (clock frequency?)

Step 7.   Open circuit file FIG 26-3. You are looking at three negative-edge-triggered J-K flip-flops wired as a 3-bit synchronous up/down binary counter. When the UP/DOWN' switch is up, the counter will count up. When the UP/DOWN' switch is down, the counter will count down. The C, D, and CLK switches should be up (1). Click the On-Off switch to run the simulation and press the C key on the computer keyboard to clear the counter. Press the C key again to raise CLR' back to binary one (1).

NOTE: If this experiment is being performed in a hardwired laboratory, use two 74112 ICs for the negative-edge-triggered J-K flip-flops, one 7408 IC for the 2-input AND gates, one 7432 IC for the 2-input OR gates, and one 7404 IC for the INVERTER.

Step 8.   With the D switch up (count up), press the space bar on the computer keyboard to make the CLK' input drop from one (1) to zero (0) to produce a negative clock edge. Record the counter binary output for one clock pulse in Table 26-2. Press the space bar on the computer keyboard enough times to produce another negative clock edge. Record the counter binary output for two clock pulses in Table 26-2. Repeat this procedure until the table is complete, then stop the simulation.

Table 26-2 Output Count

Clock Pulse	Output Q2 Q1 Q0
1	
2	
3	
4	
5	
6	
7	
8	

Question: Based on the data in Table 26-2, what conclusion can you draw about the direction of the binary count as the clock pulses are applied to the counter CLK' input? **Explain.**

Step 9.   Press the D key on the computer keyboard to make the UP/DOWN' switch go down (count down). Click the On-Off switch to run the simulation again and press the C key on the computer keyboard to clear the counter. Press the C key again to raise CLR' back to binary one (1).

Step 10.   With the D switch down, repeat the procedure in Step 8 and record the data in Table 26-3, then stop the simulation.

Table 26-3 Output Count

Clock Pulse	Output Q2 Q1 Q0
1	
2	
3	
4	
5	
6	
7	
8	

Question: Based on the data in Table 26-3, what conclusion can you draw about the direction of the binary count as the clock pulses are applied to the counter CLK′ input? **Explain.**

Step 11.   Open circuit file FIG 26-4. You will test a 74191 presettable up/down synchronous binary counter. This counter can be preset to any count by dropping the active low load input (LOAD) to binary zero (0). This will cause the binary input on the DCBA input terminals to be parallel loaded into the counter. This counter will count up one count on each positive edge of the clock pulse if the up′/down (U/D) input is set to binary zero (0). This counter will count down one count on each positive edge of the clock pulse if the up′/down (U/D) input is set to binary one (1). Notice that the UP′/DOWN switch will place a binary zero (0) on the U/D input when the **switch is up**, causing the counter to **count up**, and place a binary one (1) on the U/D input when the **switch is down**, causing it to **count down**. This counter also has an active low enable input (CTEN). The active low enable input (CTEN) must be low (0) for the counter to be enabled, otherwise it will be disabled. See the Theory section for more details about the 74191 synchronous binary counter.

Step 12.   Switches E and CLK should be down (0) and switches L and U should be up (1). Click the On-Off switch to run the simulation. Set the DCBA input to a binary five (0101). Press the L key on the computer keyboard to lower the active low load input (LOAD) to binary zero (0). Press the L key again to raise the active low load input to binary one (1).

Question: What did you observe at the counter outputs (QD–QA)?

Step 13.   With switch U up, press the space bar on the computer keyboard to produce a positive clock edge on the CLK input. Record the outputs for one clock pulse in Table 26-4. Keep pressing the space bar to produce a number of positive clock edges on the CLK input and record the outputs in Table 26-4 until the table is complete.

Table 26-4 Output Count

Clock Pulse	Output QD QC QB QA	TC	RC
1			
2			
3			
4			
5			
6			
7			
8			
9			
10			
11			
12			
13			
14			
15			
16			

Questions: What did you observe at counter outputs (QD–QA)?

What did you observe at counter outputs TC and RC?

Step 14.   Press the U key on the computer keyboard to bring down the U switch. Press the space bar on the computer keyboard to produce a positive clock edge on the CLK input. Record the outputs for one clock pulse in Table 26-5. Keep pressing the space bar to produce a number of positive clock edges on the CLK input and record the outputs in Table 26-5 until the table is complete.

Table 26-5 Output Count

Clock Pulse	Output QD QC QB QA	TC	RC
1			
2			
3			
4			
5			
6			
7			
8			
9			
10			
11			
12			
13			
14			
15			
16			

Question: What did you observe at counter outputs QD–QA?

What did you observe at counter outputs TC and RC?

Step 15.   Press the E key on the computer keyboard to raise the active low enable (CTEN) up to binary one (1). Keep pressing the space bar to apply a number of positive clock edges to the CLK input and observe the counter outputs, then stop the simulation.

Question: What did you observe at the counter outputs? **Explain.**

Step 16.   Open circuit file FIG26-5. You are looking at a circuit that will display the timing of a 74191 synchronous binary counter. Switches A, B, C, and D should be down (0). The word generator and logic analyzer settings should be as in Step 4.

> NOTE: If this experiment is being performed in a hardwired laboratory, use a pulse generator in place of the word generator. If a logic analyzer is not available, use a dual-trace oscilloscope. Connect the load input (LOAD) to +5 V and leave inputs A, B, C and D open. Skip Step 18.

Step 17.   Click the On-Off switch to run the simulation. The light blue curve plot on the logic analyzer screen is the clock input (CLK) and the first red curve plot is the parallel load input (LOAD). The dark blue (QA), dark green (QB), brown (QC), and light green (QD) curve plots represent the 4-bit counter outputs. The second red curve plot is the RCO output and the second blue curve plot is the MAX/MIN output. Draw and label the curve plots in the space provided, then stop the simulation.

Questions: What is the frequency relationship between the clock input (light blue) and the counter outputs QA (dark blue), QB (dark green), QC (brown), and QD (light green)? What is the modulus (divide-by) of this counter?

Is the counter counting up or down? **Explain why.**

Step 18.   Click the arrow in the circuit window. Change switches A, B, C, and D to binary five
(0101). This will cause the counter to start counting at the binary number five (0101). Reset
the cursor on the word generator to the top of the number column (See Step 5). Click the
On-Off switch to run the simulation again. Draw and label the new curve plots in the space
provided, then stop the simulation.

Questions: What is the first binary output (QD–QA)? Is it the same as the binary input DCBA?

Is the counter counting up or down?

At what output count does RCO go low (0)?

At what output count does MAX/MIN go high (1)?

Step 19.   Open circuit file FIG 26-6. You will test a 74191 wired as a MOD-12 (divide-by-12) counter. Notice that the MAX/MIN output and the CLK input are connected to a NAND gate that feeds the LOAD input. This will cause the counter to reset the count during the positive part of the clock period when the counter output reaches the terminal state and the MAX/MIN output is high (1). The counter will reset the binary count to the value set by the DCBA switches.

Step 20.   Click the On-Off switch to run the simulation. Press the space bar on the computer keyboard enough times to apply a series of positive clock edges to the CLK input and observe the output (QD–QA). Record the results in Table 26-6, then stop the simulation.

Table 26-6 Output Count

Clock Pulse	Output QD QC QB QA
1	
2	
3	
4	
5	
6	
7	
8	
9	
10	
11	
12	
13	
14	
15	
16	

Questions: Is the counter counting up or down?

What is the number of different binary output states for this counter?

What happened on clock pulse number thirteen? **Explain.**

Based on these results, what is the modulus (divide-by) of this counter?

How does the modulus (MOD) of this counter compare with the parallel input (DCBA) binary number?

How can you change the modulus (MOD) of this counter without changing the wiring?

Step 21.   Make sure the CLK switch is down. Change the modulus (divide-by) of this counter to
             MOD-10 without changing the wiring. Repeat Step 20. Record the results in Table 26-7,
             then stop the simulation.

Table 26-7 Output Count

Clock Pulse	Output QD QC QB QA
1	
2	
3	
4	
5	
6	
7	
8	
9	
10	
11	
12	
13	
14	
15	
16	

Questions: What is the number of different binary output states for this counter?

Based on these results, what is the modulus (divide-by) of this counter? Is it as expected?

Step 22.   Open circuit file FIG26-7. Bring down the function generator enlargement and make sure that
            the following settings are selected: *Square Wave*, Freq = 10 kHz, Duty Cycle = 50%, Ampl =
            2.5 V, Offset = 2.5 V. Move the function generator to the left. Bring down the oscilloscope
            enlargement and make sure that the following settings are selected: Time base (Scale = 200
            μs/Div, Xpos = 0, Y/T), Ch A (Scale = 5 V/Div, Ypos = 0, DC), Ch B (Scale = 5 V/Div, Ypos
            = –2, DC), Trigger (Pos. edge, Level = 1 μV, Sing, A). Move the oscilloscope to the right.
            Click the On-Off switch to run the simulation. After one full screen display, stop the
            simulation. Use the cursors on the oscilloscope to determine the time period (T) for one cycle
            and the frequency (f) of the clock input (red) and the counter output (blue). Record your
            answers in the space provided.

                    Clock input $T_c = $ _____

                                             $f_c = $ _____

                    Counter output $T_o = $ _____

                                             $f_o = $ _____

Step 23.   Based on the results in Step 22, calculate the modulus (divide-by) of this counter.

Questions: Based on the circuit configuration in Figure 26-7, what is the expected modulus (divide-
by) of this counter? **Explain why.**

How does this answer compare with the calculated modulus in Step 23?

Step 24.   Click the arrow in the circuit window. Change the modulus (divide-by) of the counter to
            MOD-8 without changing the wiring. Click the On-Off switch to run the simulation. Use
            the cursors on the oscilloscope to determine the time period (T) for one cycle and the
            frequency (f) of the clock input (red) and the counter output (blue). Record your answer in
            the space provided.

                            Clock input $T_c =$ _____

                                                       $f_c =$ _____

                            Counter output $T_o =$ _____

                                                       $f_o =$ _____

Step 25.   Based on the results in Step 24, calculate the modulus (divide-by) of this counter.

Questions: How does this calculated modulus in Step 25 compare with the expected modulus
(divide-by)?

Step 26.   Open circuit file FIG26-8. You are looking at two 74191 cascaded synchronous counters
            wired as an 8-bit counter. The counter on the right outputs the low 4-bits and the counter on
            the left outputs the high 4-bits. The cascaded 8-bit synchronous counter can count up or
            down depending on the position of the UP'/DOWN switch. Notice that the RCO output of
            the counter on the right is connected to the active low enable input (CTEN) of the counter
            on the left. This will cause the counter on the left to be enabled during the low part of the
            clock period and increment or decrement one count when the counter on the right reaches
            its terminal count and begins a new count cycle.

Step 27.   Switch U should be up and the CLK switch should be down (0). Click the On-Off switch to
            run the simulation. Press the space bar on the computer keyboard enough times to apply a
            series of positive clock edges to the CLK input to make the counter count more than sixteen
            counts and observe the output, then stop the simulation.

Questions: Is the counter counting up or down?

What is the maximum count of this counter in binary and decimal?

What is the modulus (MOD) of this counter?

Step 28.   Press the U key on the computer keyboard to lower the UP'/DOWN switch. Make sure the CLK switch is up (1). Click the On-Off switch to run the simulation. Press the space bar on the computer keyboard enough times to apply a series of positive clock edges to the CLK input to make the counter count more than sixteen counts and observe the output, then stop the simulation.

Question: Is the counter counting up or down? **Explain why.**

EXPERIMENT

27 BCD Counters

Objectives:

1. Demonstrate how J-K flip-flops are used to build asynchronous BCD counters.
2. Demonstrate the operation of a 74290 asynchronous decade counter as a BCD counter.
3. Investigate 74290 asynchronous BCD counter timing.
4. Demonstrate how J-K flip-flops are used to build synchronous BCD counters.
5. Demonstrate the operation of the 74190 presettable synchronous decade counter as a BCD counter.
6. Investigate 74190 synchronous BCD counter timing.
7. Demonstrate cascaded BCD counters.
8. Demonstrate a BCD counter display system using 7-segment displays.

Materials:

One 5 V dc power supply
Eight logic switches
Ten logic probe lights
Four negative-edge-triggered J-K flip-flops (2-74112 ICs)
One asynchronous decade counter (1-74290 IC)
Two presettable synchronous decade counters (2-74190 ICs)
Four two-input AND gates (1-7408 IC)
One two-input OR gate (1-7432 IC)
One INVERTER (1-7404 IC)
Two BCD-to-7-segment decoder drivers (2-7447 ICs)
Two common anode 7-segment LED displays
One function (pulse) generator
One logic analyzer or dual-trace oscilloscope
Fourteen 300 Ω resistors

Theory:

Make sure you complete Experiments 25 and 26 before attempting this experiment. Also, review the Theory sections of those experiments.

Counters that have **ten output states (MOD-10)** are often referred to as **decade counters**, regardless of the sequence of the count. Decade counters are often used for dividing a pulse frequency by ten. A decade counter with a count sequence of zero (0000) through nine (1001) is commonly called a **BCD counter** because its ten output states consist of the BCD code. BCD counters find widespread use in applications where pulses or events are to be counted with the results displayed on a decimal readout.

In order to build a BCD counter, it is necessary to force a 4-bit counter to reset (clear) before completing all of its normal output states. A BCD counter must reset (clear) to binary zero (0000) on the tenth count after reaching binary nine (1001). Therefore, it will count between zero (0000) and nine (1001). One way to make a binary counter reset (clear) after the count of nine (1001) is to decode count ten (1010) with a NAND gate and use the output of the NAND gate to reset (clear) the counter. This will cause a **glitch** on the Q1 output just before the counter resets (clears). This glitch can be eliminated by using logic circuitry to control the timing of each counter flip-flop instead of using the reset (clear) counter input to clear the counter. This requires a more complex circuit and raises the cost of the counter IC chip.

The circuit in Figure 27-1 consists of four **negative-edge-triggered J-K flip-flops** wired as an **asynchronous BCD counter** using logic circuitry to control the timing of each counter flip-flop. This counter is asynchronous because all of the flip-flops are not clocked simultaneously. Because the flip-flop J and K inputs are controlled by different logical outputs to force each flip-flop to toggle at the correct time for a zero (0000) to nine (1001) count, this counter will not produce any glitches in the output pulses. The counter **active low CLR′ input** is formed by connecting all of the flip-flop active low CLR inputs together. When the counter CLR′ input is dropped to binary zero (0), all of the flip-flops will clear (0), causing the counter to be cleared. The counter CLR′ input must be returned to binary one (1) in order for the counter to count. If the active low counter CLR′ input is not returned to binary one (1), the counter will stay cleared because the flip-flop CLR inputs override the flip-flop CLK inputs. This circuit is similar to the circuit in the popular 74290 (7490) **asynchronous decade counter**, except the output of the first flip-flop (Q0 or QA) is not connected internally to the clock input of the second flip-flop in the 74290 (7490).

The 74290 in Figure 27-2 is an **asynchronous decade counter**. It is wired internally as a **1-bit MOD-2 (divide-by-2) counter** with a negative-edge-triggered clock input (INA) and output QA, and a **3-bit MOD-5 (divide-by-5) counter** with a negative-edge-triggered clock input (INB) and outputs QB, QC, and QD. THE 74290 in Figure 27-2 is externally wired as a BCD counter by connecting the output of the 1-bit MOD-2 counter (QA) to the clock input of the 3-bit MOD-5 counter (INB). This feature makes the 74290 very versatile because it can be wired as a 1-bit (MOD-2), 3-bit (MOD-5), or **4-bit (MOD-10) counter**. The 74290 also has two **active high reset (clear) inputs (R01 and R02)** and two **active high set inputs (R91 and R92)**. The reset (clear) and set inputs can be used to reset the counter or cut off the binary count in order to change the modulus (divide-by) of the counter. By wiring the correct combination of counter outputs to the reset (clear) inputs (R01 and R02) or the set inputs (R91 and R92), the modulus (divide-by) of the counter can be changed to any value between MOD-2 (divide-by-2) and MOD-10 (divide-by-10).

The circuit in Figure 27-3 will display the **timing** of the **74290 asynchronous decade counter** configured as a BCD counter by connecting output QA to INB. The word generator will apply the clock pulses to the counter clock input (INA). The logic analyzer will monitor the clock pulses and the counter outputs (QA–QD). The active high reset (clear) inputs (R01 and R02) are set to binary one (1) by the word generator during the first clock pulse to clear the counter, and then return to zero for the remaining clock pulses. The active high set inputs (R91 and R92) are permanently connected to ground (0) to allow the counter to count.

The circuit in Figure 27-4 consists of four **negative-edge-triggered J-K flip-flops** wired as a **synchronous BCD counter**. This is a synchronous counter because all of the flip-flop clock inputs are connected to the counter CLK′ input and are clocked simultaneously. Each flip-flop J and K input is connected to either 5 V (J = 1, K = 1), or the output of a logic network. The correct logic causes the flip-flops to toggle at the proper time to produce a zero (0000) to nine (1001) count. The counter **active low CLR' input** is formed by connecting all of the flip-flop active low CLR inputs together. When the counter CLR' input is dropped to binary zero (0), all of the flip-flops will clear (0), causing the counter to be cleared. The counter CLR' input must be returned to binary one (1) in order for the counter to count. Because this synchronous counter has more extensive circuitry than the asynchronous counter, it is **more expensive** to build but it can be clocked at a **higher clock frequency**.

The 74190 in Figure 27-5 is a **presettable synchronous decade counter**. It can be preset to any count by applying a 4-bit binary number to inputs DCBA and dropping the **active low load input (LOAD)** to binary zero (0). The load input (LOAD) must be returned to binary one (1) in order for the counter to count because the load input overrides the CLK input. The 74190 will count up if a binary zero (0) is applied to the U'/D input and it will count down if a binary one (1) is applied to the U'/D input. The counter will increment or decrement one count each time the CLK input receives a **positive clock edge**. The **active low count enable input (CTEN)** must be low (0) for the counter to be enabled, otherwise it will be disabled. The **terminal count (MAX/MIN) output** is normally low (0) but will go high (1) for one clock period when the counter reaches zero (0000) in the count down mode or nine (1001) in the count up mode. The **ripple clock (RCO) output** is normally high (1), but will go low (0) when the counter is zero (0000) in the count down mode or nine (1001) in the count up mode. The MAX/MIN and RCO outputs are made available in order to be able to cascade 74190 synchronous decade counters to provide a count higher than BCD nine (1001). Each additional cascaded counter adds a decimal digit to the output.

The circuit in Figure 27-6 will display the **timing** of the **74190 synchronous decade counter**. The word generator will apply the clock pulses to the counter clock input. The logic analyzer will monitor the clock pulses, the counter outputs (QA– QD), the ripple clock output (RCO), and the terminal count output (MAX/MIN).

The circuit in Figure 27-7 consists of two 74190 **cascaded synchronous BCD counters**. The counter on the right outputs the **least significant BCD digit** and the counter on the left outputs the **most significant BCD digit**. This cascaded synchronous BCD counter can count up or down depending on the position of the UP'/DOWN switch. Notice that the RCO output of the counter on the right is connected to the active low enable input (CTEN) of the counter on the left. This will cause the counter on the left to be enabled and increment or decrement one count when the counter on the right reaches its terminal count and begins a new count cycle. Because both counters are clocked simultaneously, this cascaded pair is **synchronous**.

The circuit in Figure 27-8 consists of two 74190 synchronous BCD counters connected to two BCD-to-7-segment decoder/drivers driving LED displays. The 7-segment LED displays will output the decimal equivalent of the BCD output code from each of the BCD counters. The right display will output the **least significant decimal digit** and the left display will output the **most significant decimal digit**. The CLK logic switch will provide a positive clock edge to clock both synchronous BCD counters simultaneously, making the display system synchronous.

Figure 27-1 Asynchronous BCD Counter

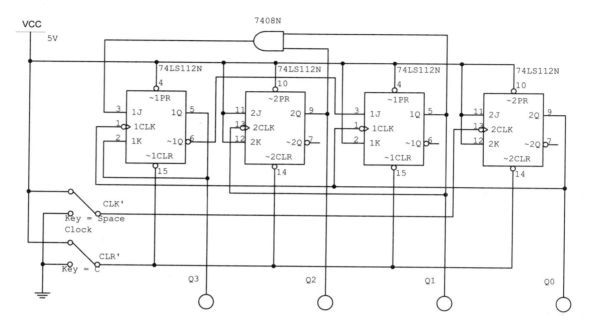

Figure 27-2 74290 Asynchronous BCD Counter

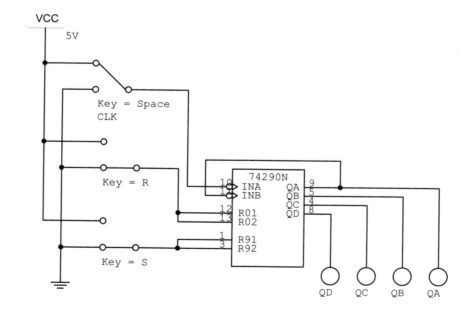

Figure 27-3 74290 Asynchronous BCD Counter Timing

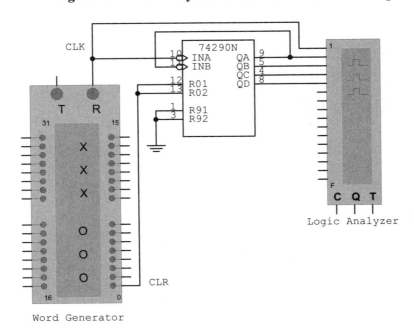

Figure 27-4 Synchronous BCD Counter

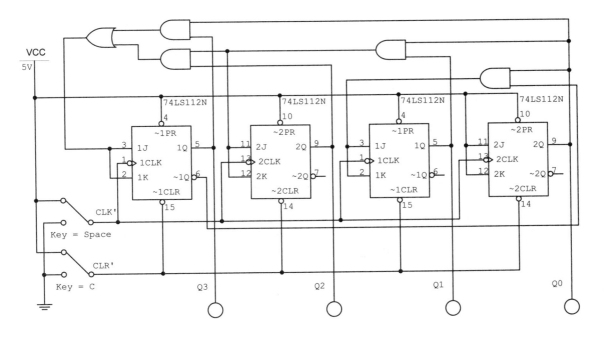

Figure 27-5 74190 Presettable Synchronous Decade Counter

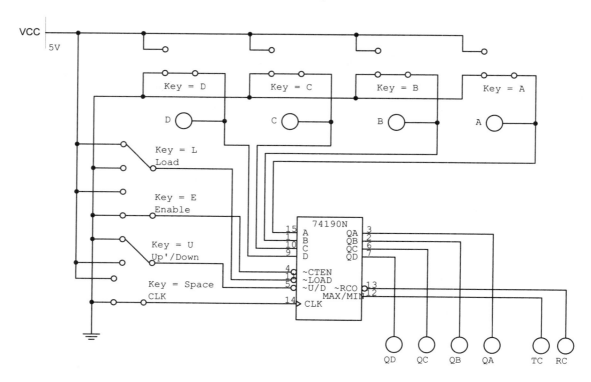

Figure 27-6 74190 Synchronous Decade Counter Timing

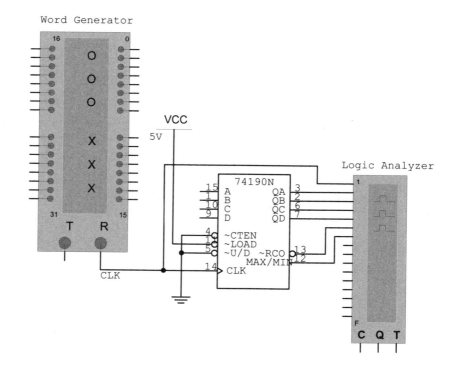

Experiment 27   295

Figure 27-7 Cascaded 74190 Synchronous BCD Counters

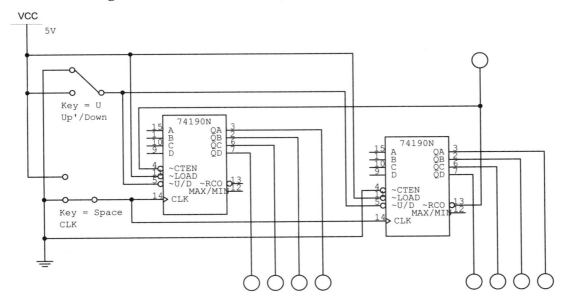

Figure 27-8 BCD Counter Display System

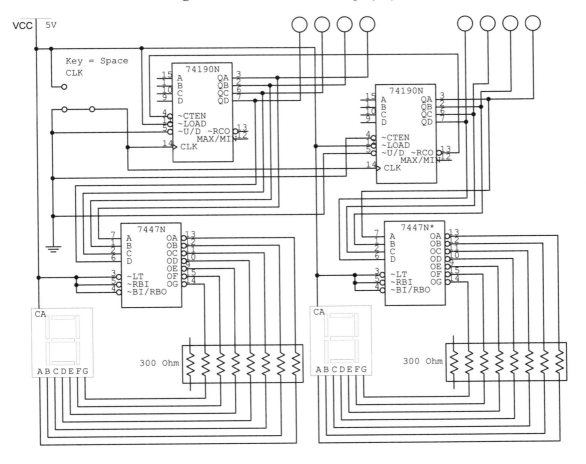

Procedure:

Step 1.   Open circuit file FIG27-1. You are looking at four negative-edge-triggered J-K flip-flops wired as an asynchronous BCD counter. The C and CLK switches should be up (1). Click the On-Off switch to run the simulation and press the C key on the computer keyboard to clear the counter. Press the C key again to raise the CLR' back to binary one (1).

Questions: Is the clear input (CLR') active low or active high?

Why does the CLR' input need to be binary one (1) in order for the counter to be able to count?

Step 2.   Press the space bar on the computer keyboard to make the CLK' input drop from one (1) to zero (0) to produce a negative clock edge. Record the counter binary output for one clock pulse in Table 27-1.

Table 27-1 Output Count

Clock Pulse	Output $Q3\ Q2\ Q1\ Q0$
1	
2	
3	
4	
5	
6	
7	
8	
9	
10	
11	
12	
13	
14	
15	
16	

Question: Is the counter positive-edge-triggered or negative-edge-triggered?

Step 3.   Press the space bar on the computer keyboard enough times to produce another negative clock edge. Record the counter binary output for two clock pulses in Table 27-1. Repeat this procedure until the table is complete, then stop the simulation.

Questions: Based on the data in Table 27-1, what conclusions can you draw about the relationship between the counter binary output and the number of clock pulses applied to the counter CLK′ input?

What was the maximum count?

What happened when the counter was clocked after the maximum count was reached?

Step 4.   Open circuit file FIG27-2. You will test a 74290 asynchronous decade counter. This counter has two active high reset (clear) inputs (R01 and R02), two active high set inputs (R91 and R92), and two negative-edge-triggered clock inputs (INA and INB). The INA input is used to trigger flip-flop A and the INB input is used to trigger the 3-bit MOD-5 counter using flip-flops B, C, and D. The 74290 in Figure 27-2 is externally wired as a BCD counter by connecting flip-flop output QA to clock input INB and using INA as the clock input. See the Theory section for more details about the 74290 asynchronous decade counter.

Step 5.   Switches R and S should be down (0) and the CLK switch should be up (1). Click the On-Off switch to run the simulation. If the counter output (QD–QA) is not cleared, press the R key on the computer keyboard to raise the reset (clear) input (R01 and R02) to binary one (1) to clear the counter. Press the R key again to lower the reset input back to binary zero (0).

Step 6.   Press the space bar on the computer keyboard to produce a negative clock edge on the INA input. Keep pressing the space bar to produce a number of negative clock edges on the INA input and cycle the counter through its full count range.

Question: What did you observe at counter outputs QD–QA?

Step 7.   Press the S key on the computer keyboard and observe the counter output (QD–QA), then stop the simulation.

Question: What did you observe at counter outputs QD–QA?

Step 8.   Open circuit file FIG27-3. You are looking at a circuit that will display the timing of a 74290 asynchronous decade counter, configured as a BCD counter by connecting output QA to INB and using INA as the clock input. Bring down the word generator enlargement and make sure that the following settings are selected: Frequency = 1 kHz, Trigger = Internal, Controls = Burst, Hex codes (00000001 once, then 00000000 repeated thirty-one times). Move the word generator to the left. Bring down the logic analyzer enlargement and make sure that the following settings are selected: Clocks/Div = 32, Clock Setup (Clock Source = Internal, Clock Rate = 10 kHz, Pre-trigger Samples = 100, Post-trigger Samples = 1000, Threshold Voltage = 2.5), Trigger Settings (Trigger Clock Edge = Positive, Trigger Qualifier = x, Pattern A= xxxxxxxxxxxxxxxx, Trigger Combinations = A). Move the logic analyzer to the right.

NOTE: If this experiment is being performed in a hardwired laboratory, use a pulse generator in place of the word generator. If a logic analyzer is not available, use a dual-trace oscilloscope.

Step 9.   Click the On-Off switch to run the simulation. The light blue curve plot on the logic analyzer screen is the clock input (INA). The dark blue (QA), the dark green (QB), brown (QC), and light green (QD) curve plots represent the BCD counter outputs. Draw and label the curve plots in the space provided, then stop the simulation.

Questions: Are the output binary counts correct after each pulse (negative edge)?

What happened on the tenth clock pulse (negative edge)?

What is the frequency relationship between the INA input (light blue) and counter output QD (light green)? What is the modulus (divide-by) of this counter?

Step 10.   Open circuit file FIG 27-4. You are looking at four negative-edge-triggered J-K flip-flops wired as a synchronous BCD counter. The C and CLK switches should be up (1). Click the On-Off switch to run the simulation and press the C key on the computer keyboard to clear the counter. Press the C key again to raise CLR' back to binary one (1).

> NOTE: If this experiment is being performed in a hardwired laboratory, use two 74112 ICs for the negative-edge-triggered J-K flip-flops, one 7408 IC for the 2-input AND gates, and one 7432 IC for the 2-input OR gate.

Step 11.   Press the space bar on the computer keyboard to make the CLK' input drop from one (1) to zero (0) to produce a negative clock edge. Record the counter binary output for one clock pulse in Table 27-2.

Table 27-2 Output Count

Clock Pulse	Output Q3 Q2 Q1 Q0
1	
2	
3	
4	
5	
6	
7	
8	
9	
10	
11	
12	
13	
14	
15	
16	

Question: Is the counter positive-edge-triggered or negative-edge-triggered?

Step 12.   Press the space bar on the computer keyboard enough times to produce another negative clock edge. Record the counter binary output for two clock pulses in Table 27-2. Repeat this procedure until the table is complete, then stop the simulation.

Questions: Based on the data in Table 27-2, what conclusion can you draw about the relationship between the counter binary output and the number of clock pulses applied to the counter CLK' input?

What was the maximum count?

What happened when the counter was clocked after the maximum count was reached?

How did these outputs compare with the output of the asynchronous counter in Table 27-1?

Step 13.   Open circuit file FIG 27-5. You will test a 74190 presettable synchronous decade counter. This counter can be preset to any count by dropping the active low LOAD input to binary zero (0). This will cause the binary input on the DCBA input terminals to be parallel loaded into the counter. This counter will count up one count on each positive edge of the clock pulse if binary zero (0) is applied to the up'/down (U'/D) input, and count down one count on each positive edge of the clock pulse if binary one (1) is applied to the up'/down (U'/D) input. Notice that the UP'/DOWN switch will place a binary zero (0) on the U'/D input when the switch is up, causing the counter to count up. This counter also has an active low enable input (CTEN), which must be low (0) for the counter to be enabled, otherwise it will be disabled. See the Theory section for more details about the 74190 synchronous decade counter.

Step 14.   Switches E and CLK should be down (0) and switches L and U should be up (1). Click the On-Off switch to run the simulation. Make sure the DCBA input is set to binary zero (0000). If the counter output (QD, QC, QB, QA) is not cleared, press the L key on the computer keyboard to lower the active low LOAD input to binary zero (0) and load all zeros into the counter. Press the L key again to raise the active low LOAD input to binary one (1).

Step 15.   With switch U up, press the space bar on the computer keyboard to produce a positive clock edge on the CLK input. Record the outputs for one clock pulse in Table 27-3. Keep pressing the space bar to produce a number of positive clock edges on the CLK input and record the output in Table 27-3 until the table is complete, then stop the simulation.

Table 27-3 Output Count

Clock Pulse	Output QD QC QB QA	TC	RC
1			
2			
3			
4			
5			
6			
7			
8			
9			
10			
11			
12			
13			
14			
15			
16			

Questions: What did you observe at counter outputs QD–QA?

What was the maximum count?

What happened when the counter was clocked after the maximum count was reached?

What did you observe at counter outputs TC and RC?

Step 16.   Click the On-Off switch to run the simulation again. Set the DCBA switches to a binary nine (1001). Press the L key to load the counter with nine (1001). Press the L key again to raise the LOAD input back to binary one (1). Press the U key to bring down the U switch. Press the space bar on the computer keyboard to produce a **positive clock edge** on the CLK input. Record the outputs for one clock pulse in Table 27-4. Keep pressing the space bar to produce a number of positive clock edges on the CLK input and record the outputs in Table 27-4 until the table is complete, then stop the simulation.

Table 27-4 Output Count

Clock Pulse	Output QD QC QB QA	TC	RC
1			
2			
3			
4			
5			
6			
7			
8			
9			
10			
11			
12			
13			
14			
15			
16			

Questions: What did you observe at counter outputs QD–QA?

What did you observe at counter outputs TC and RC?

Step 17.   Click the On-Off switch to run the simulation again. Press the E key on the computer
            keyboard to raise the active low enable (CTEN) to binary one (1). Keep pressing the space
            bar to apply a number of positive clock edges to the CLK input and observe the counter
            outputs, then stop the simulation.

Question: What did you observe at the counter outputs? **Explain.**

Step 18.   Open circuit file FIG27-6. You are looking at a circuit that will display the timing of a
            74190 synchronous decade counter. The word generator and logic analyzer settings should
            be as in Step 8.

NOTE: If this experiment is being performed in a hardwired laboratory, use a pulse generator in
place of the word generator. If a logic analyzer is not available, use a dual-trace oscilloscope.

Step 19.   Click the On-Off switch to run the simulation. The light blue curve plot on the logic
            analyzer screen is the clock input (CLK). The dark blue (QA), dark green (QB), brown
            (QC), and light green (QD) curve plots represent the counter outputs. The red curve plot is
            the RCO output and the second dark blue curve plot is the MAX/MIN output. Draw and
            label the curve plots in the space provided, then stop the simulation.

Questions: Is the counter counting up or down? **Explain why.**

Are the output binary counts correct after each clock pulse (positive edge)?

What did you observe at outputs RCO and MAX/MIN?

What is the frequency relationship between the clock input (light blue) and counter output QD (light green)? What is the modulus (divide-by) of this counter?

Step 20.   Open circuit file FIG27-7. You are looking at two cascaded 74190 synchronous BCD counters. The counter on the right outputs the least significant BCD digit and the counter on the left outputs the most significant BCD digit. This cascaded synchronous BCD counter can count up or down depending on the position of the UP'/DOWN switch. Notice that the RCO output of the counter on the right is connected to the active low enable input (CTEN) of the counter on the left. This will cause the counter on the left to be enabled and increment or decrement one count when the counter on the right reaches its terminal count and begins a new count cycle.

Step 21.   Switch U should be up and the CLK switch should be down (0). Click the On-Off switch to run the simulation. Press the space bar on the computer keyboard enough times to apply a series of positive clock edges to the CLK input to make the counter count beyond twenty counts and observe the output, then stop the simulation.

Questions: Is the counter counting in BCD or binary?

Is the counter counting up or down?

What is the maximum count for each counter?

What is the maximum count for this cascaded counter?

Step 22.   Press the U key on the computer keyboard to lower the UP'/DOWN switch. Make sure the CLK switch is up (1). Click the On-Off switch to run the simulation. Press the space bar on the computer keyboard enough times to apply a series of positive clock edges to the CLK input to make the counter count beyond twenty counts and observe the output, then stop the simulation.

Question: Is the counter counting in BCD or binary?

Is the counter counting up or down? **Explain why.**

Step 23.   Open circuit file FIG27-8. You are looking at two cascaded 74190 synchronous BCD counters connected to two BCD-to-7-segment decoder/drivers driving two LED displays. The 7-segment displays will output the decimal equivalent of the BCD output code for each of the BCD counters. The clock switch should be down (0). See the Theory section for more details.

NOTE: It is recommended that this step be performed in a hardwired laboratory to obtain experience wiring and testing an actual logic circuit. Use two 7447 BCD-to-7-segment decoder/drivers and two common anode LED displays with 300 Ω resistors, as shown in Figure 27-8.

Step 24.   Click the On-Off switch to run the simulation. Press the space bar on the computer
            keyboard enough times to apply a series of positive clock edges to the CLK input to make
            the counter count at least twenty counts and observe the output, then stop the simulation.

NOTE: Wait for the 7-segment display to change to the next number before applying another
clock edge.

Questions: What is the highest count for each BCD counter?

What happened after the counter on the right reached its highest count?

What is the highest count of the cascaded pair?

What would be required to produce a higher count?

Did the BCD output from each BCD counter match the decimal output on the LED displays?

EXPERIMENT

28 Troubleshooting Sequential Logic Circuits

Objectives:

1. Determine the defective flip-flop in a 4-bit asynchronous (ripple) counter wired using four J-K flip-flops.
2. Determine the defective component in a 4-bit synchronous counter wired using four J-K flip-flops and two 2-input AND gates.
3. Determine the defective counter in a cascaded counter pair.
4. Determine the correct output frequency for a 74191 presettable synchronous counter wired as a frequency divider.
5. Determine if the output waveshapes are correct for a particular 74191 presettable synchronous counter circuit configuration.

Materials:

This experiment can only be performed on Electronics Workbench Multisim using the circuits disk provided with this manual.

Theory:

In order to perform these experiments effectively, you must first complete Experiments 21–27. Also review the Theory sections of those experiments before beginning these troubleshooting exercises. Use the theory learned in Experiments 21–27 to answer the questions or find the defective component in the logic circuits in these experiments.

1.   Open circuit file FIG28-1. You are looking at a 4-bit **asynchronous** (ripple) counter wired
      using four J-K flip-flops. Double-click the logic analyzer to bring down the enlargement. Click
      the On-Off switch to run the simulation. Based on the waveshapes on the logic analyzer screen,
      determine which flip-flop (Q0, Q1, Q2, or Q3) is defective. Explain how you determined your
      answer.

      Defective flip-flop _____

2.   Open circuit file FIG28-2. You are looking at a 4-bit **synchronous** (ripple) counter wired using
      four J-K flip-flops and two 2-input AND gates. Double-click the logic analyzer to bring down
      the enlargement. Click the On-Off switch to run the simulation. Based on the waveshapes on
      the logic analyzer screen, determine which component (flip-flops Q0, Q1, Q2, or Q3 or AND
      gates G1A or G1B) is defective. Explain how you determined your answer.

      Defective component _____

3.   Open circuit file FIG28-3. You are looking at a 4-bit **synchronous** counter wired using four J-K flip-flops and two 2-input AND gates. Double-click the logic analyzer to bring down the enlargement. Click the On-Off switch to run the simulation. Based on the waveshapes on the logic analyzer screen, determine which component (flip-flops Q0, Q1, Q2, or Q3 or AND gates G1A or G1B) is defective. Explain how you determined your answer.

        Defective component _____

4.   Open circuit file FIG28-4. You are looking at a 4-bit **synchronous** counter wired using four J-K flip-flops and two 2-input AND gates. Double-click the logic analyzer to bring down the enlargement. Click the On-Off switch to run the simulation. Based on the waveshapes on the logic analyzer screen, determine which component (flip-flops Q0, Q1, Q2, or Q3 or AND gates G1A or G1B) is defective. Explain how you determined your answer.

        Defective component _____

5.   Open circuit file FIG28-5. You are looking at cascaded counters. Double click the oscilloscope to bring down the enlargement. Click the On-Off switch to run the simulation. Use the cursors on the oscilloscope screen to determine the output time period (T) and the output frequency (f) for each counter (C1 and C2). Change the oscilloscope Timebase scale, if necessary. Record your answers in the space provided.

        $T_1 =$ _____ $f_1 =$ _____

        $T_2 =$ _____ $f_2 =$ _____

Question: Based on the wiring configuration of each counter and the input frequency of 60 kHz, are the output frequencies correct? If not, which counter (C1 or C2) is defective?

6.   Open circuit file FIG28-6. Double click the oscilloscope to bring down the enlargement. Click the On-Off switch to run the simulation. Use the cursors on the oscilloscope screen to determine the output (blue curve plot) time period (T) and the output frequency (f) for the counter. Record your answers in the space provided.

$$T = \underline{\hspace{3cm}} \qquad\qquad f = \underline{\hspace{3cm}}$$

Question: Based on the circuit configuration in Figure 28-6, is the 74191 counter producing the correct output frequency? If not, what should the frequency be? **Why?**

7.   Open circuit file FIG28-7. Double click the logic analyzer to bring down the enlargement. Click the On-Off switch to run the simulation.

Question: Are the output waveshapes correct for the 74191 presettable synchronous counter circuit configuration in Figure 28-7? If not, what is wrong with the output waveshapes?

PART

V Interfacing the Analog World

n the experiments in Part V, you will learn how the **digital world** is interfaced with the **analog world**. Most real-world physical variables are **analog** in nature, which means they have values that are within a continuous range. On the other hand, **digital** systems are **discrete** in nature, which means they have values that are **in steps**. Any **analog data** that must be input into a digital system must first be converted to **digital form**. Any **digital data** that must be output into the real analog world must first be converted nto **analog form**. A circuit that converts digital codes into analog data is called a **digital-to-analog converter (DAC)**. A circuit that converts analog data into digital codes is called an **analog-to-digital converter (ADC)**. Digital-to-analog converters and analog-to-digital converters will be studied in the first wo experiments in Part V.

There are many applications in which analog data must be converted into digital codes and transferred into a digital system. This process is called **data acquisition**. The principles of data acquisition will be studied in the last experiment in Part V.

> **If the experiments in Part V are performed in a hardwired laboratory, make sure you save the D/A converter circuit wired in Experiment 29 and the A/D converter circuit wired in Experiment 30. They will be used in the data acquisition experiment (Experiment 31).**

The circuits for the experiments in Part V can be found on the enclosed disk in the PART5 subdirectory.

EXPERIMENT

29

Digital-to-Analog Converters

Objectives:

1. Demonstrate the relationship between the digital input and the analog output of a D/A converter (DAC).
2. Demonstrate how to set the full-scale output (range) of a D/A converter.
3. Demonstrate how to measure the output offset of a D/A converter.
4. Develop an understanding of the concept of resolution as applied to D/A converters.
5. Demonstrate how a staircase output can be used to test a D/A converter and measure its resolution.

Materials:

One dc voltage supply (+5 V, +12 V, −12 V)
Eight logic switches
Eight logic probe lights
One D/A converter (1-1408 or DAC0808 IC)
One op-amp (1-741 IC)
Two synchronous counters (2-74191 ICs)
One 0–2 kΩ potentiometer
One 0.1 μF capacitor
Two 1 kΩ resistors
One 0–10 V voltmeter
One function (pulse) generator
One oscilloscope

Theory:

Most real-world physical variables are **analog** in nature and have values that are within a **continuous** range. On the other hand, **digital systems** are **discrete** in nature and have values that are one of two possibilities. Any **digital data** that must be output into the real analog world must first be converted into **analog form**. A circuit that converts digital codes into analog data is called a **digital-to-analog converter (DAC)**. The output of a D/A converter (DAC) can be in the form of a **voltage** or a **current**.

A D/A converter (DAC) will convert a **digital binary input** to an **output voltage or current** that is proportional to the magnitude of the binary input value. The **DAC full scale-output** is the output value when all ones are applied to the DAC binary input. The full-scale output value determines the **range** of the DAC.

The **DAC output offset** is the value of the output when all binary zeros are applied to the DAC binary input. In an ideal DAC, the output offset is zero. In a real DAC, the output offset is not zero. Many DACs have an external offset adjustment that allows you to zero the output offset.

The **resolution** of a DAC is defined as the smallest change that can occur in the analog output as a result of a change in the digital input. The DAC resolution (step size) can be measured by measuring the change in the output voltage or current for a **single-step change** in the binary input. The **calculated DAC resolution (step size)** is equal to the full-scale output divided by the number of input steps from zero to full-scale output. The **number of input steps** depends solely on the number of DAC input bits, where a 4-bit DAC has 15 steps and an 8-bit DAC has 255 steps (0 is not a step). The **percent resolution** is equal to the percentage of the full-scale output that one step represents. Therefore, the percentage resolution can be calculated by dividing the step size by the full-scale output times 100%, which is equal to the inverse of the number of steps times 100%. This means that the **larger the number of bits**, the higher the number of steps and the **smaller the resolution (step size)**. For this reason, most manufacturers usually specify a DAC resolution as the number of bits.

The 8-bit voltage output DAC circuit in Figure 29-1a or Figure 29-1b will help develop an understanding of the relationship between the digital input and the analog output of a DAC. The DAC full-scale output voltage is set by first applying all ones (11111111) to the DAC binary input, and then adjusting the potentiometer to the output voltage desired for a full-scale value.

The **1408 (DAC0808) 8-bit D/A converter** in Figure 29-1b (hardwired circuit) consists of an R/2R ladder network and produces an output current that is proportional to the magnitude of the binary input. The **maximum output current** (full-scale output current) is determined by dividing 5 V by the value of resistor R. With the value of R shown in Figure 29-1b, the full-scale output current is 5 mA (5 V/1 kΩ). This is the value of the output current (I_o) when all ones (11111111) are applied to the DAC binary input. In order to convert this output current to an output voltage, the **741 op-amp current-to-voltage converter** circuit is required. The **output voltage (V_o)** is equal to the DAC output current (I_o) multiplied by the value of resistor R_F. With all ones applied to the DAC binary input, the full-scale output voltage can be adjusted to any desired value to a maximum of 10 V (5 mA × 2 kΩ) by adjusting the value of resistor R_F. The value of the resistor connected to V_{REF}^- should be equal to the value of resistor R connected to V_{REF}^+ to balance the DAC.

The circuit in Figure 29-2 is an 8-bit voltage output DAC with its binary input connected to a word generator set up as an 8-bit counter and an oscilloscope monitoring the output. This 8-bit count will cause the 8-bit DAC to cover the full output range in 255 steps. You will notice that the oscilloscope output will look more like a straight line (more like an analog output) than a staircase. The larger the number of steps, the smaller the step size and the more closely the DAC output will represent a true analog output (better resolution) for any given voltage range. This circuit can be used to measure the DAC resolution or test the DAC to determine if any steps are missing by expanding the oscilloscope horizontal and vertical scales and examining the staircase steps.

Figure 29-1a Digital-to-Analog Conversion

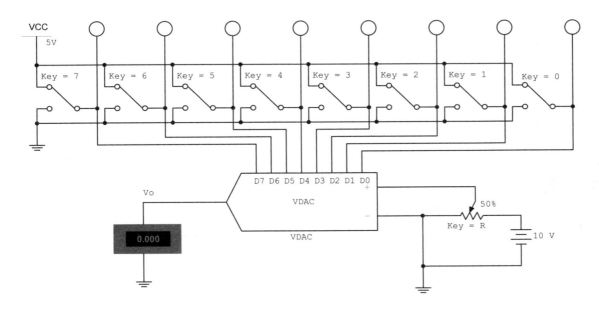

Figure 29-1b Digital-to-Analog Conversion—Hardwired Circuit

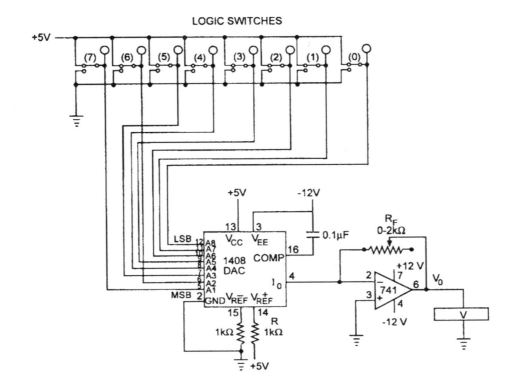

Figure 29-2 DAC Output Waveform (Staircase)

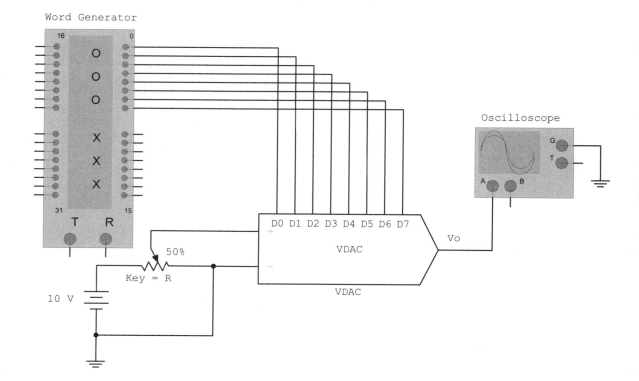

Procedure:

Step 1.   Open circuit file FIG29-1. You are looking at an 8-bit voltage output DAC circuit that will help develop an understanding of the relationship between the digital input and the analog output. The 0–2 kΩ potentiometer is used to set the DAC full-scale output voltage (voltage range). All of the input logic switches should be up (1).

> NOTE: It is recommended that this experiment be performed in a hardwired laboratory. If a hardwired laboratory is available, wire the DAC circuit in Figure 29-1b using a 1408 (DAC0808) current output DAC and a 741 op-amp current-to-voltage converter circuit.

Step 2.   With all of the logic switches up (1) click the On-Off switch to run the simulation. Adjust the 0–2 kΩ potentiometer until the DAC output voltage is as close as possible to 5 V. This will set the DAC to a full-scale output voltage of 5 V.

Step 3.   By pressing the 0–7 keys on the computer keyboard, change the DAC 8-bit input to binary zero (00000000). Record the DAC output voltage (V_o) for a binary zero input in Table 29-1.

Table 29-1 DAC Output

Binary Input	V_o (V)
0000 0000	
0000 0001	
0000 0010	
0000 0100	
0000 1000	
1111 1111	

Step 4.   Press the 0 key on the computer keyboard to change the DAC 8-bit input to binary one (00000001). Record the DAC output voltage for a binary one input in Table 29-1. Change the logic switches to the remaining binary input values in Table 29-1 and record the DAC output voltages, then stop the simulation.

Questions: Based on the data in Table 29-1, what is the DAC output offset voltage?

Based on the data in Table 29-1, what is the DAC full-scale output voltage (voltage range)?

Based on the data in Table 29-1, what is the DAC resolution? (step size)?

Based on the data in Table 29-1, is the DAC output voltage proportional to the magnitude of the binary input?

Step 5.   Based on the DAC full-scale output voltage and the number of steps for an 8-bit input, calculate the expected resolution (step size) of the DAC in Figure 29-1.

Question: How did your calculated resolution (step size) in Step 5 compare with the measured value in Step 4?

Step 6.   Based on the data in Table 29-1, calculate the DAC percent resolution.

Step 7.   Open circuit file FIG29-2. You are looking at an 8-bit voltage output DAC with its binary
              input connected to a word generator set up as an 8-bit counter that will produce a
              continuous binary output count from 0 (00000000) to 255 (11111111). This will cause the
              8-bit DAC to cover the full 0–5 V output range in 255 steps. Bring down the word
              generator enlargement and make sure that the following settings are selected: Frequency =
              1 kHz, Trigger = Internal, Controls = Cycle, Hex codes (00000000 to 000000FF). Move the
              word generator to the left. Bring down the oscilloscope enlargement and make sure that the
              following settings are selected: Time base (Scale = 50 ms/Div, Xpos = 0, Y/T), Ch A (Scale =
              2 V/Div, Ypos = 0, DC), Ch B (Scale = 5 V/Div, Ypos = 0, DC), Trigger (Pos. edge, Level = 1
              µV, Sing, A). Move the oscilloscope to the right. Click the On-Off switch to run the simulation.
              **The oscilloscope curve plot may take a long time to develop, depending on your computer
              speed.** After a half screen display, stop the simulation. Notice how closely the curve plot on the
              oscilloscope screen resembles a straight line instead of a "staircase" waveform.

NOTE: If this experiment is being performed in a hardwired laboratory, modify the circuit in
Figure 29-1b by replacing the logic switches with two 74191 4-bit counters wired as an 8-bit
counter, as shown in Figure 26-8. Use a function generator 1 kHz pulse output to provide the clock
input to the 8-bit counter. Replace the voltmeter connected to the output (V_o) with an oscilloscope.

Questions: Why does the DAC output curve plot on the oscilloscope screen look more like a straight
line than a "staircase"?

Based on the oscilloscope curve plot, what is the full-scale output voltage of the DAC?

Step 8.   Change the oscilloscope Timebase scale to 1 ms/Div and the Channel A scale to 50 mV/Div.
              The DAC output shown on the oscilloscope screen is called a "staircase" waveform. Use the
              oscilloscope cursors to measure the DAC resolution (step size) and record your answer in the
              space provided.

                        Resolution (step size) = _____

Question: How did your measured resolution (step size) compare with the value determined in Step 5?

NOTE: Don't forget to save this circuit for Experiment 31, if you are using a hardwired circuit.

EXPERIMENT

30

Analog-to-Digital Converters

Objectives:

1. Demonstrate the relationship between the analog input and the digital output of a A/D converter (ADC).
2. Demonstrate how to set the input voltage range of an A/D converter.
3. Investigate ADC conversion and demonstrate how to start conversion.
4. Develop an understanding of the concept of resolution (quantization error) as applied to A/D converters.
5. Demonstrate how to wire a continuous running A/D converter.

Materials:

One 5 V dc voltage supply
One normally open pushbutton switch
Eight logic probe lights
One A/D converter (1-ADC0804 IC)
Two 0–2 kΩ potentiometers
One 15 pF capacitor
One 10 kΩ resistors
Two 0–10 V voltmeters

Theory:

Most real-world physical variables are **analog** in nature and have values that are within a continuous range. On the other hand, **digital** systems are **discrete** in nature and have values that are one of two possibilities. Any **analog data** that must be input into a digital system must first be converted into digital form. A circuit that converts analog voltages into digital codes is called an **analog-to-digital converter (ADC)**. The magnitude of the digital output is proportional to the analog input voltage. The **ADC full-scale input** is equal to the analog input voltage that produces the highest binary output (all binary ones). The full-scale input voltage determines the **input range** of the ADC.

The A/D conversion process is more complex and time consuming than the D/A conversion process because A/D converters require a **conversion time** before a digital output is obtained. Several important types of A/D converters utilize a **D/A converter (DAC)** and a **comparator** as part of their circuitry. The two most common types are the **digital ramp ADC (counter-type)** and the **successive approximation ADC**. In the digital ramp ADC, a binary counter output is connected to a DAC input. The DAC output voltage is compared to the analog input voltage by the **comparator**. When the counter reaches a binary number that causes the DAC output voltage to equal the analog input voltage, the counter stops counting and conversion is complete. The counter binary output is used as the ADC digital output. In the successive approximation ADC, the counter is replaced by a **register** that is connected to the DAC binary input. Control logic successively modifies the register output until the DAC output voltage is equal to the analog input voltage and conversion is complete. The register binary output is used as the ADC digital output. The **successive approximation ADC** has a **shorter conversion time** than the digital ramp ADC. The **flash ADC** has the **shortest conversion time**, but it requires much more logic circuitry, making it more expensive and complicated to build. Advancements in integrated circuit technology are reducing the cost of the flash ADC.

The **resolution (quantization error)** of an ADC is defined as the largest analog input voltage change that can occur without producing a change in the digital output. The ADC resolution (quantization error) can be calculated by dividing the full-scale analog input voltage by the number of digital output steps from zero to the maximum count (all binary ones). The number of digital output steps depends solely on the number of ADC output bits, where a 4-bit ADC has 15 output steps and an 8-bit ADC has 255 output steps (0 is not a step). This means that the larger the number of bits, the higher the number of steps and the smaller the resolution (quantization error).

The 8-bit ADC circuit in Figure 30-1a or Figure 30-1b will be used to demonstrate the relationship between the analog input and the digital output for an A/D converter. The **ADC full-scale input voltage** (input voltage range) is set by adjusting the 0–2 kΩ RANGE potentiometer to the reference voltage (VREF) desired for full-scale input. The **analog input voltage (V_{IN})** is set by adjusting the 0–2 kΩ INPUT potentiometer. The input voltage (V_{IN}) can be varied between 0 V and 5 V. In Figure 30-1a, ADC conversion is started by starting the simulation. (The START CONV switch is not needed for the computer simulation.) When conversion is complete (the digital output is ready), the end-of-conversion (EOC) output will rise to binary one. Because the end-of-conversion (EOC) output is connected to the start-of-conversion (SOC) input, a new conversion cycle will be immediately started after each end-of-conversion. This will cause the ADC to operate continuously, making the output respond immediately to any changes in the analog input. The active high output enable (OE) is used to enable the ADC digital output. In Figure 30-1a, it is connected to 5 V (1) to continuously enable the ADC output. The end-of-conversion output (EOC) and the output enable (OE) are used for timing when interfacing the ADC to a microprocessor or computer data bus.

The **hardwired ADC0804** in Figure 30-1b is an **8-bit successive approximation ADC** that is designed to be interfaced with a microprocessor. It has an **internal clock** with a **frequency** determined by the value of **R and C**. The **frequency of the clock** controls the length of the **ADC conversion time**. The ADC0804 has a **differential analog input** $\left(V_{IN}^{+} \text{ and } V_{IN}^{-} \right)$ which can be used to apply a differential input voltage. The differential input is wired as a **single-ended input** by connecting V_{IN}^{-} to ground. The ADC0804 requires a **reference voltage** that is **1/2 the voltage desired for a full-scale input**. If the $V_{REF/2}$ terminal is left open, the full-scale input voltage is equal to V_{cc} (5 V). The ADC0804 also has a digital ground (D_{GND}) and an analog ground (A_{GND}) so that the analog and digital circuits can be isolated from each other. To **start conversion**, the active low **start-of-conversion input (SOC)** must be pulsed to binary zero for a short duration. When conversion is complete (the digital output is ready), the active low **end-of-conversion (EOC) output** will drop to binary zero. If the EOC output is connected to the SOC input, the ADC will operate continuously, as described previously. The ADC0804 has an **active low output enable (RD)** and an **active low chip select (CS)** for interfacing to a microprocessor. In Figure 30-1b, these inputs are connected to ground so that the ADC output is continuously enabled.

Figure 30-1a   Analog-to-Digital Conversion

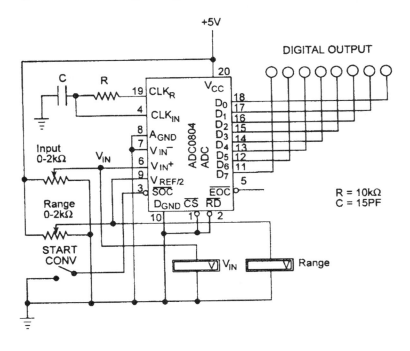

Figure 30-1b   Analog-to-Digital Conversion—Hardwired Circuit

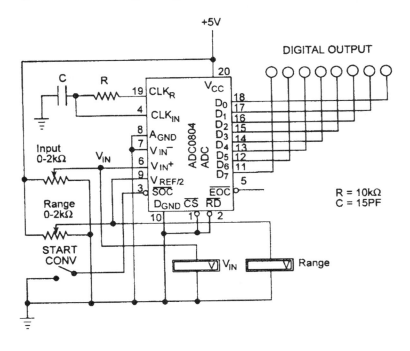

Procedure:

Step 1.   Open circuit file FIG30-1. You are looking at an 8-bit A/D converter that will be used to demonstrate the relationship between the analog input and the digital output. The 0–2 kΩ RANGE potentiometer is used to set the ADC full-scale input voltage (input voltage range). The 0–2 kΩ INPUT potentiometer is used to vary the analog input voltage between 0 V and 5 V. The end-of-conversion (EOC) output is connected to the start-of-conversion (SOC) input for a continuous running ADC.

> NOTE: It is recommended that this experiment be performed in a hardwired laboratory. If a hardwired laboratory is available, wire the ADC circuit in Figure 30-1b using an ADC0804 A/D converter. Connect $\overline{EOC}$ output (pin 5) to $\overline{SOC}$ input (pin 3) for a continuous running ADC.

Step 2.   The START CONV switch should be up for the hardwired circuit in Figure 30-1b. (The START CONV switch is not used in the computer simulation. The start of the simulation automatically starts conversion.) Click the On-Off switch to run the simulation or turn the power on if using a hardwired circuit. Adjust the 0–2 kΩ RANGE potentiometer until the RANGE voltage is as close as possible to 5 V (2.5 V for the hardwired ADC in Figure 30-1b). This will set the ADC to a full-scale input voltage of 5 V (input voltage range of 0–5 V). The potentiometer setting can be changed by typing capital R or small r on the computer keyboard, changing the % setting.

Step 3.   Adjust the 0–2 kΩ INPUT potentiometer to a V_{IN} as close as possible to 1.0 V. (Press the START CONV switch to start conversion if using the hardwired circuit, then return the switch to open.) Record the ADC digital output for an analog input voltage (V_{IN}) of 1.0 V in Table 30-1. Repeat the procedure for the remaining values of V_{IN} in Table 30-1 and record the digital outputs, then stop the simulation. If the digital output does not change when V_{IN} changes, restart the simulation to start conversion again.

Table 30-1 ADC Binary Output

V_{IN} (V)	Digital Output	Decimal Equiv.
0		
1.0		
2.0		
3.0		
4.0		
5.0		

Step 4.   Calculate the decimal equivalent of each binary output in Table 30-1 and record your answers.

Questions: Based on the data in Table 30-1, what is the ADC full-scale input voltage (input voltage range)?

Based on the data in Table 30-1, is the ADC digital output magnitude proportional to the magnitude of the analog input voltage?

Step 5.   Based on the data in Table30-1, calculate the resolution (quantization error) of the ADC in Figure 30-1.

Question: Based on the quantization error, how much can the analog input voltage (V_{IN}) be varied before the digital output changes by one count?

Step 6.   Start the simulation to start conversion again. (In the hardwired circuit, turn on the circuit and press and release the START CONV switch to start conversion.) Adjust the 0–2 kΩ RANGE potentiometer until the RANGE voltage is as close as possible to 2.5 V (1.25 V for the hardwired ADC in Figure 30-1b). Adjust the 0–2 kΩ INPUT potentiometer to vary V_{IN} from zero until the digital output just reaches maximum (11111111), then stop the simulation. Record the analog input voltage for maximum digital output in the space provided.

                        Analog input voltage = _____

Question: Based on the result in Step 6, what is the new ADC full-scale input voltage (input voltage range?) Is it what you expected?

NOTE: Don't forget to save this circuit for Experiment 31, if you are using a hardwired circuit.

EXPERIMENT

31

Data Acquisition

Objectives:

1. Demonstrate how the quality of the analog data being acquired is affected by the relationship between the frequency of the analog data and the sampling rate of the data acquisition system..
2. Demonstrate how the conversion time of the A/D converter in a data acquisition system affects the input data sampling rate.

Materials:

One D/A converter circuit wired in Experiment 29 (Figure 29-1b modified).
One A/D converter circuit wired in Experiment 30 (Figure 30-1b modified).
One function generator
One dual-trace oscilloscope

Theory:

The process of storing analog data as digital data is called **data acquisition**. In order to store an analog voltage waveshape (voltages that are changing in value with time), an A/D converter must sample the analog voltage frequently so that important voltage values will not be missed. This requires a **continuously running ADC** that starts conversion immediately after the previous conversion is complete. When the analog data is a fast-changing high-frequency waveshape, the data sampling points must be close together. This requires an ADC with a **short conversion time**. If the conversion time is too long (**sampling rate too low**) for the frequency of the analog waveshape being acquired, the **analog reproduction** of the digitally stored waveshape will be **distorted**. The shorter the conversion time (higher the sampling rate), the lower the distortion of the analog reproduction.

The circuit in Figure 31-1 will be used to demonstrate how the relationship between ADC conversion time (sampling rate) and the frequency of the analog input voltage affects the distortion of the analog reproduction of a digital conversion. The digital output of the A/D converter studied in Experiment 30 (Figure 30-1a or Figure 30-1b modified) is connected to the digital input of the D/A converter studied in Experiment 29 (Figure 29-1a or Figure 29-1b modified). The ADC end-of-conversion output (EOC) is connected to the ADC start-of-conversion input (SOC) to start conversion immediately after the previous conversion is complete. This will make the ADC run continuously. The **conversion time** of the ADC will determine the **time between sampling points (sampling rate)**. The analog potentiometer has been replaced by a function generator. The function generator will apply a periodic sine wave to the ADC analog input. The dual trace oscilloscope will monitor the A/D converter analog input and the D/A converter analog output. The relationship between the ADC conversion time (sampling rate) and the frequency of the periodic sine wave will determine how accurately the D/A converter analog output reproduces the A/D converter analog input. If the conversion time (sampling rate) is not fast enough, the D/A converter analog output will not be a good representation of the A/D converter analog input.

Figure 31-1 Data Acquisition

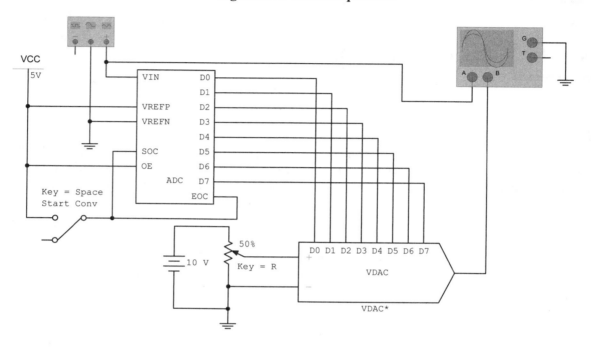

Procedure:

Step 1.   Open circuit file FIG31-1. The digital output of the A/D converter studied in Experiment
            30 (Figure 30-1a) is connected to the digital input of the D/A converter studied in
            Experiment 29 (Figure 29-1a). The ADC end-of-conversion output (EOC) is connected to
            the ADC start-of-conversion input (SOC) to start conversion immediately after the previous
            conversion is complete, causing the ADC to run continuously. **In the computer
            simulation, conversion will start automatically when the analysis begins**. The
            voltmeters, the logic probe lights, and the ADC range adjustment potentiometer have been
            removed and the ADC VREFP has been connected to 5 V. The analog voltage input
            potentiometer has been replaced by a function generator. The function generator will apply
            a periodic sine wave to the ADC analog input. The dual trace oscilloscope will monitor the
            A/D converter analog input and the D/A converter analog output. Bring down the function
            generator enlargement and make sure that the following settings are selected: *Sine Wave*, Freq
            = 10 kHz, Ampl = 2 V, Offset = 2 V. Move the function generator to the left. Bring down the
            oscilloscope enlargement and make sure that the following settings are selected: Time base
            (Scale = 20 µs/Div, Xpos = 0, Y/T), Ch A (Scale = 2 V/Div, Ypos = −2, DC), Ch B (Scale = 2
            V/Div, Ypos = 0, DC), Trigger (Pos. edge, Level = 1 µV, Sing, A). Move the oscilloscope to
            the right.

NOTE FOR HARDWIRED LABORATORY USERS: It is recommended that this experiment be performed in a hardwired laboratory. You will use the ADC wired in Experiment 30 (Figure 30-1b modified) and the DAC wired in Experiment 29 (Figure 29-1b modified). Make sure that the ADC and the DAC are adjusted for a 5 V full-scale. You will connect the digital output of the ADC to the digital input of the DAC. The ADC end-of-conversion output (EOC) is connected to the ADC start-of-conversion input (SOC) to start conversion immediately after the previous conversion is complete, causing the ADC to run continuously. The START CONV switch should remain in the circuit because **conversion needs to be initially started in the hardwired circuit**. The dc voltmeter that measures V_{IN} should be removed. The logic probe lights should be removed. The input potentiometer should be replaced by a function generator. The function generator will apply a 0–4 V periodic sine wave to the ADC analog input. A dual-trace oscilloscope should monitor the A/D converter analog input and the D/A converter analog output, as shown in Figure 31-1. The function generator frequency should be 1 kHz and the oscilloscope time base should be 200μs/div for the hardwired circuit.

Step 2.   Click the On-Off switch to run the simulation. After one full screen display, stop the simulation. In the computer simulation, conversion will start automatically. (In the hardwired circuit, the START CONV switch must be pushed to start conversion. Make sure you return the switch to its original position after the conversion starts.) Notice the ADC 10 kHz (1 kHz for the hardwired circuit) periodic input sine wave (red curve plot) and the DAC output (blue curve plot) on the oscilloscope.

Question: Is the DAC output sine wave (blue) a good representation of the ADC input sine wave (red)?

Step 3.   Change the frequency of the function generator to 50 kHz (5 kHz for the hardwired circuit). Change the oscilloscope time base to 5 μs/div (50 μs/div for the hardwired circuit). Click the On-Off switch to run the simulation again. After one full screen display, stop the simulation. Notice the ADC 50 kHz (5 kHz for the hardwired circuit) periodic input sine wave (red curve plot) and the DAC output (blue curve plot) on the oscilloscope.

Question: Is the DAC output sine wave (blue) as good a representation of the ADC input sine wave (red) as when the input frequency was 10 kHz (1 kHz for the hardwired circuit)? **If not, why not?**

Step 4.   Use the cursors on the oscilloscope to measure the time between sampling points and
            record your answers in the space provided.

            Time = _____

Question: Based on your answer in Step 4, what is the conversion time of the A/D converter?

IC Chip Pin Diagrams

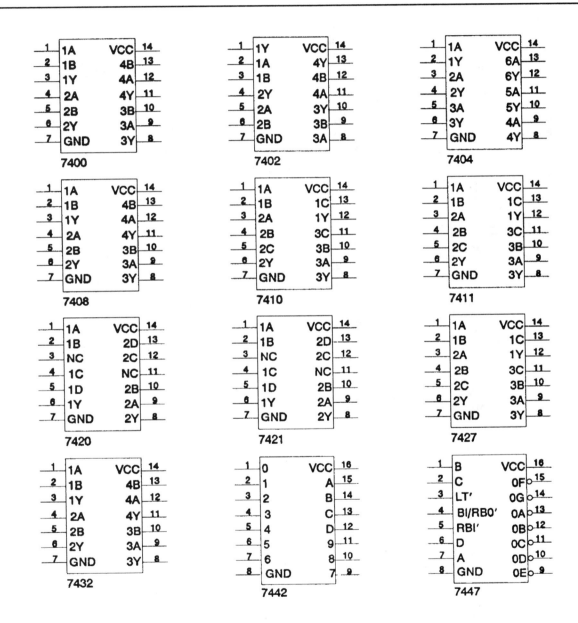

7400
1	1A	VCC	14
2	1B	4B	13
3	1Y	4A	12
4	2A	4Y	11
5	2B	3B	10
6	2Y	3A	9
7	GND	3Y	8

7402
1	1Y	VCC	14
2	1A	4Y	13
3	1B	4B	12
4	2Y	4A	11
5	2A	3Y	10
6	2B	3B	9
7	GND	3A	8

7404
1	1A	VCC	14
2	1Y	6A	13
3	2A	6Y	12
4	2Y	5A	11
5	3A	5Y	10
6	3Y	4A	9
7	GND	4Y	8

7408
1	1A	VCC	14
2	1B	4B	13
3	1Y	4A	12
4	2A	4Y	11
5	2B	3B	10
6	2Y	3A	9
7	GND	3Y	8

7410
1	1A	VCC	14
2	1B	1C	13
3	2A	1Y	12
4	2B	3C	11
5	2C	3B	10
6	2Y	3A	9
7	GND	3Y	8

7411
1	1A	VCC	14
2	1B	1C	13
3	2A	1Y	12
4	2B	3C	11
5	2C	3B	10
6	2Y	3A	9
7	GND	3Y	8

7420
1	1A	VCC	14
2	1B	2D	13
3	NC	2C	12
4	1C	NC	11
5	1D	2B	10
6	1Y	2A	9
7	GND	2Y	8

7421
1	1A	VCC	14
2	1B	2D	13
3	NC	2C	12
4	1C	NC	11
5	1D	2B	10
6	1Y	2A	9
7	GND	2Y	8

7427
1	1A	VCC	14
2	1B	1C	13
3	2A	1Y	12
4	2B	3C	11
5	2C	3B	10
6	2Y	3A	9
7	GND	3Y	8

7432
1	1A	VCC	14
2	1B	4B	13
3	1Y	4A	12
4	2A	4Y	11
5	2B	3B	10
6	2Y	3A	9
7	GND	3Y	8

7442
1	0	VCC	16
2	1	A	15
3	2	B	14
4	3	C	13
5	4	D	12
6	5	9	11
7	6	8	10
8	GND	7	9

7447
1	B	VCC	16
2	C	0F	15
3	LT'	0G	14
4	BI/RB0'	0A	13
5	RBI'	0B	12
6	D	0C	11
7	A	0D	10
8	GND	0E	9

7474

1	1CLR'	VCC	14
2	1D	2CLR'	13
3	1CLK	2D	12
4	1PRE'	2CLK	11
5	1Q	2PRE'	10
6	1Q'	2Q	9
7	GND	2Q'	8

7475

1	1Q'	1Q	16
2	1D	2Q	15
3	2D	2Q'	14
4	3C,4C	1C,2C	13
5	VCC	GND	12
6	3D	3Q'	11
7	4D	3Q	10
8	4Q'	4Q	9

7483

1	A4	B4	16
2	Σ3	Σ4	15
3	A3	C4	14
4	B3	C0	13
5	VCC	GND	12
6	Σ2	B1	11
7	B2	A1	10
8	A2	Σ1	9

7486

1	1A	VCC	14
2	1B	4B	13
3	1Y	4A	12
4	2A	4Y	11
5	2B	3B	10
6	2Y	3A	9
7	GND	3Y	8

74290

1	R91	VCC	14
2	NC	R02	13
3	R92	R01	12
4	QC	INB	11
5	QB	INA	10
6	NC	QA	9
7	GND	QD	8

74293

1	NC	VCC	14
2	NC	R02	13
3	NC	R01	12
4	QC	INB	11
5	QB	INA	10
6	NC	QA	9
7	GND	QD	8

74112

1	1CLK	VCC	16
2	1K	1CLR'	15
3	1J	2CLR'	14
4	1PRE'	2CLK	13
5	1Q	2K	12
6	1Q'	2J	11
7	2Q'	2PRE'	10
8	GND	2Q	9

74121

1	$\overline{Q}$	VCC	14
2	NC	NC	13
3	A1	NC	12
4	A2	Rext/Cext	11
5	B	Cext	10
6	Q	Rint	9
7	GND	NC	8

74138

1	A	VCC	16
2	B	Y0	15
3	C	Y1	14
4	G2A'	Y2	13
5	G2B'	Y3	12
6	G1	Y4	11
7	Y7	Y5	10
8	GND	Y6	9

74147

1	4	VCC	16
2	5	NC	15
3	6	D	14
4	7	3	13
5	8	2	12
6	C	1	11
7	B	9	10
8	GND	A	9

74148

1	4	VCC	16
2	5	E0	15
3	6	GS	14
4	7	3	13
5	E1	2	12
6	A2	1	11
7	A1	0	10
8	GND	A0	9

74151

1	D3	VCC	16
2	D2	D4	15
3	D1	D5	14
4	D0	D6	13
5	Y	D7	12
6	W	A	11
7	G'	B	10
8	GND	C	9

74157

1	A/B'	VCC	16
2	1A	G'	15
3	1B	4A	14
4	1Y	4B	13
5	2A	4Y	12
6	2B	3A	11
7	2Y	3B	10
8	GND	3Y	9

74173

1	M	VCC	16
2	N	CLR	15
3	1Q	1D	14
4	2Q	2D	13
5	3Q	3D	12
6	4Q	4D	11
7	CLK	G2'	10
8	GND	G1'	9

74190

1	B	VCC	16
2	QB	A	15
3	QA	CLK	14
4	CTEN'	RCO'	13
5	D/U'	MAX/MIN	12
6	QC	LOAD'	11
7	QD	C	10
8	GND	D	9

74191

```
1  B          VCC  16
2  QB         A    15
3  QA         CLK  14
4  CTEN'      RCO' 13
5  D/U'   MAX/MIN  12
6  QC         LOAD' 11
7  QD         C    10
8  GND        D     9
```
74191

74194

```
1  CLR'   VCC  16
2  SR     QA   15
3  A      QB   14
4  B      QC   13
5  C      QD   12
6  D      CLK  11
7  SL     S1   10
8  GND    S0    9
```
74194

74280

```
1  G      VCC  14
2  H      F    13
3  NC     E    12
4  I      D    11
5  EVEN   C    10
6  ODD    B     9
7  GND    A     8
```
74280

555

```
1  GND    VCC  8
2  TRI    DIS  7
3  OUT    THR  6
4  RES    CON  5
```

B | Notes on Using Electronics Workbench Multisim

1. If you wish to remove a component from a circuit, disconnect both terminals from the circuit; otherwise you may get an error message.

2. You can change a component value by double clicking it with the mouse and changing the menu value using the keyboard.

3. You can bring down an instrument enlargement by double clicking the instrument with the mouse.

4. The circuit disk provided with this manual is write protected; therefore, you cannot save a changed circuit to the disk. If you wish to save a changed circuit, you must select "Save As" in the File menu and save it on another disk or the hard drive.

5. The color of a logic analyzer or oscilloscope curve trace is the same as the color of the circuit wire connected to the input. A wire color can be changed by right clicking the wire with the mouse and selecting a color from the table on the screen.

6. Wires can be moved by placing the arrow on the wire, clicking the left mouse button, and dragging the wire to a new position.

Bibliography

Floyd, T. L. *Digital Fundamentals.* 8th ed. Upper Saddle River, NJ: Prentice Hall, 2003.

Klietz, W. *Digital Electronics: A Practical Approach.* 6th ed. Upper Saddle River, NJ: Prentice Hall, 2002

Leach, D. P., and Malvino, A. P. *Digital Principles and Applications.* 5th ed. New York: McGraw-Hill, 1995

Tocci, R. J., and Widmer, N. S. *Digital Systems: Principles and Applications.* 9th ed. Upper Saddle River, NJ: Prentice Hall, 2004